"Dr. Stanley draws on research from l  other leading scientists to share brea[ ] forces and materials are at work inside [ ] fect their surfaces and atmospheres. She also sheds light on the mysteries of our beautiful—but often fierce—solar system, leaving us with a deeper appreciation of the preciousness of life on our home planet."

—Lisa J. Graumlich, PhD; President, American Geophysical
Union (AGU); Dean Emeritus, College of the Environment,
University of Washington

"*What's Hidden Inside Planets?* is a great read that's filled with the wonder and joy of scientific discoveries connected to the unseen Earth underfoot, the planetary menagerie in our solar system, and beyond. Sabine Stanley's insights also afford a window into discovery-based science itself—how we 'know what we know' about both near and distant worlds—providing a fabulous 'on-ramp' linking what planetary scientists do to what they've discovered."

—C.M. Bailey, PhD; Professor and Chair, Department of Geology,
College of William & Mary; President, Geological Society of America

"A mesmerizing glimpse into the inner secrets of our own planet and its siblings. With wit and humor, Sabine Stanley plays planetary decoder, helping us reimagine Earth and the other planets as continually evolving under the sway of the universe's implacable forces. I couldn't put it down."

—Alanna Mitchell, PhD, Science journalist, playwright,
and author of *The Spinning Magnet: The Electromagnetic
Force that Created the Modern World and Could Destroy It*

"Dr. Stanley's chronicle of humanity's planetary explorations is an exciting and approachable guide to what makes planets tick, inside and out!"

—Jeff Coughlin, PhD, PMP; Astrophysicist, SETI Institute;
K2 Science Office Director, NASA Kepler Mission

"A charming and quirky meander through the interiors of the planets around us and through a life in science. Come for the deep knowledge of how planets operate, but stay for the delightful asides about *Friends*, the unexpected pleasures of white chocolate in pâté, and why Venus is a terrible planet."

—Mike Brown, PhD; Director, Caltech Center for Comparative Planetary
Evolution; author of *How I Killed Pluto and Why It Had It Coming*

"Sabine Stanley serves up a sumptuous smörgåsbord of science, diving into the deep puzzles of planets and moons. A slathering of kitchen analogies, and a sprinkling of delicious tidbits about the author's evolution, make the physics easy to digest. A treat!"

—Ralph Lorenz, PhD; Planetary Scientist and Author, Johns Hopkins
Applied Physics Lab; Dragonfly Mission Architect and Participating Scientist
in NASA and international missions to Mars and Venus

"Although the popular media often portrays science as a sterile world of equations, *What's Hidden Inside Planets?* is a down-to-earth reminder that research and discovery are ultimately human experiences. Weaving in anecdotes from her own life, Dr. Stanley reveals how we're intimately tied to the unseen processes churning beneath our feet."

—Catherine Neish, PhD; Associate Professor of Earth Sciences,
Western University; Senior Scientist, Planetary Science Institute

What's Hidden Inside Planets?

JOHNS HOPKINS
WAVELENGTHS

In classrooms, field stations, and laboratories in Baltimore and around the world, the Bloomberg Distinguished Professors of Johns Hopkins University are opening the boundaries of our understanding of many of the world's most complex challenges. The Johns Hopkins Wavelengths series brings readers inside their stories, presenting the pioneering discoveries and innovations that benefit people in their neighborhoods and across the globe in artificial intelligence, cancer research, epidemiology, food systems, health equity, marine robotics, planetary science, science diplomacy, and other critical areas of study. Through these compelling narratives, their insights will spark conversations from dorm rooms to dining rooms to boardrooms.

This multimedia science communication program is a partnership between the Johns Hopkins University Press and the University's Office of Research. Team members include:

Consultant Editor: John Wenz
Senior Acquisitions Editor: Tiffany Gasbarrini
Copyeditor: Charles Dibble
Art Director: Molly Seamans
Designer: Matthew Cole
Interior Illustrations and Graphics: Nicole Kit
Production Manager: Jennifer Paulson
Publicists: Core Four Media and Nardi Media
Director of Strategic Engagement: Anna Marlis Burgard
JHUP Executive Director: Barbara Kline Pope
Assistant Vice Provost for Research: Julie Messersmith

What's Hidden
Inside Planets?

SABINE STANLEY, PHD

With John Wenz

Johns Hopkins University Press
Baltimore

Johns Hopkins Wavelengths is a registered trademark of the Johns Hopkins University

Johns Hopkins University Press
2715 North Charles Street
Baltimore, Maryland 21218
www.press.jhu.edu

Cataloging-in-Publication Data is available from the Library of Congress.
A catalog record for this book is available from the British Library.

ISBN: 978-1-4214-4816-9 (paperback)
ISBN: 978-1-4214-4817-6 (ebook)
ISBN: 978-1-4214-4921-0 (ebook, open access)

Special discounts are available for bulk purchases of this book. For more information, please contact Special Sales at specialsales@jh.edu.

Capturing the Cover Image

When Michael Meyers had the opportunity in April 2018 to photograph the Halema'uma'u Crater on Hawaii's Big Island, its lava levels were exceptionally high, which added to the drama of the imagery and the vibrance of his experience. He didn't know that just a week later, the volcano would erupt—and reshape part of the island. In his words:

Some photographic situations create next-level scales of intensity. When I was in front of the crater and beneath the stars that night, I imagined the amount of power and energy it takes to turn millions of tons of rock into a liquid, or the unfathomable amount of time it took to form that vast arm of the Milky Way. But what amazed, and amazes, me most is how beautiful it is all on its own. No one painted or sculpted this volcano or the Milky Way—or anything else in the universe—to get it to look that way. There is nothing artificial. To me, no work of art can compare with the beauty of the night sky or some of the natural features of our planet, and the fact that they're nothing more than the physical manifestation of the laws that govern our universe makes me love them and marvel at them even more. When I see things like this, I just try my best to capture them in a way that other people can feel what it was like for me when I was standing there, including a little bit of "Holy shit, I can't believe I'm seeing this right now!" Sometimes nature gives you a real gift—something you could never expect or imagine; this was definitely one of those moments.

Contents

ix   Preface

1   **CHAPTER 1:** Gazing Inward

29   **CHAPTER 2:** Gazing Outward

66   **CHAPTER 3:** Telltale Planetary Parcels

91   **CHAPTER 4:** Fierce and Formative Forces

120   **CHAPTER 5:** How We Peer Inside Planets

148   **CHAPTER 6:** Curious Planetary Elements

190   **CHAPTER 7:** The Future of Planetary Exploration

217   Acknowledgments

221   Notes

231   Index

Preface

THIS IS HOW I LIKE TO PICTURE IT: Like in the opening of *Bambi*, tranquil music filled the air, flowers bloomed, and animals with fluffy tails dotted the landscape. But then I remember that this was 1.8 billion years ago in the Proterozoic eon when the first continents started to stabilize, oxygen began building up in the atmosphere, and the most complex life forms around, microorganisms, were just starting to photosynthesize. So, a quick reset: Imagine an idyllic setting for these microbes; perhaps they're basking in the sun in shallow waters, soaking up some rays and adding more oxygen to the atmosphere. Suddenly, a giant asteroid or comet—similar to the one that would kill the dinosaurs more than 1.7 billion years later—smashes into the ground (which, at a time before fungi made their way to land, paving the way for plants, was an inhospitable landscape of rocks), incinerating everything in an eighty-mile-wide radius (sorry, microbes), and forcing vast amounts of debris to hurtle out in parabolic arcs in every direction, with some rock fragments landing as far as five hundred miles away. The impact was so explosive that it liquified mile-deep rock beneath the bottom of the crater, creating a roughly thirty-mile-wide melt pool of rocks and metals.

This spot, 1.8 billion years later, was under my feet while I was growing up, beneath the soil of Sudbury, Ontario, Canada. I didn't know it at the time, but it's a perfect place to reflect on the intriguing and volatile interior worlds of planets.

The Sudbury Crater is the third largest in the world, just a bit smaller than the Chicxulub Crater in Mexico left in the wake of the asteroid impact that contributed to the demise of the dinosaurs. Its impact brought materials up from the depths of Earth, creating a melt pool of magma filled with metals like nickel, copper, and gold that we use for making stainless steel, electrical wiring, jewelry, and a host of other tools and decorative objects. In 1883, the fortuitous discovery of rusty rock by blacksmith Thomas Flanagan kicked off modern-day mining in Sudbury, making it one of Earth's major mining centers— although First Nations people had extracted these resources ten thousand years earlier.

Sudbury is a geologist's dream landscape. Awesome rocks are everywhere. Like, *everywhere*. In the early 1970s, Apollo 16 and 17 astronauts even trained in Sudbury to recognize the kind of rock formations created by impacts. One rock type is called *breccia* because it's made from broken fragments of other rocks that are cemented together, often after a meteorite impact. There's a video of Apollo 16 astronaut John Young hopping up to a boulder on the Moon and proclaiming, "It looks like a Sudbury breccia, and that's the truth!"

In retrospect, it should have been obvious to me that I was supposed to become a planetary scientist. It wasn't. Everything in my environment screamed "PLANET!" but I was stubborn and didn't listen well. I was raised in a community whose entire reason for existing was that a giant space rock had smashed into it a while back. But I didn't really understand any of that when I was young, not even that all the mining was happening because of an impact. I just knew that a lot of my classmates' parents were miners.

My parents weren't miners; instead, they fed them and many others in their family-style Italian restaurant called Trevi. I never learned to cook there, but it stirred my first musings about what I wanted to do as a profession. While sitting in the small office waiting for my mom to finish work so I could go home, I'd stare at the plastic desk mat with all sorts of business cards secured below the transparent film (clearly this was before iPhones and contacts lists), and reading business cards of people with jobs like "baker," "restaurant equipment supplier," and "radio station manager" got me thinking about what I might want to be someday.

I got lots of work experience at the restaurant. I was a dishwasher, table-setter, hostess, and then, once old enough, a bartender or part of the wait staff. Staying in the family business certainly would have made my parents happy, but it became very clear that I was not well suited to it. Let's just say if you tried to order coffee or soup, I would spend several minutes trying to convince you to choose something else, because there was zero chance I was going to make it back to your table with a hot cup or bowl of anything without most of it spilled all over my tray. I tried a stint in the kitchen once, but I couldn't even make french fries properly. Even today, my partner is the chef. My ineptitude in cooking and handling liquids was perhaps the first sign that I wasn't destined to be a chef—or a laboratory-based scientist.

Despite excelling in math and science, I didn't know any scientists or university professors, so neither career path crossed my mind as a profession. My parents assumed that I'd become a doctor—which came true, just not in the medical sense. But back in grade school, I wanted to be a news anchor. They seemed like the smartest people because they had so much to tell us. In my parents' office, staring into an imaginary camera, I'd read anything out loud—including a plaque with Hallmark-like sentiments, clearly a gift from someone who had no idea what to get my parents:

> I've been in this business a long time.
> I've seen a lot. I've heard a lot.
> I've been shouted at, spit at, stared at,
> hung up on, held up, and robbed.
> The only reason I'm staying in this business
> is to see what's going to happen next!

I would then close with a quarter-turn of my desk chair to stare into the other imaginary camera and say in a news anchor voice, "Back to you, Dan." I didn't become the next Barbara Walters, but it instilled in me a love of communicating information—which would come in handy when, as an adult, I'd speak to rooms of hundreds of people about the science of planetary interiors, or when I'd write this very book for the public. My eight-year-old self would be pleased.

I went to an all-girls Catholic high school, called Marymount College. Like public schools, Catholic schools in Canada were free, and about half the schools in my hometown were run by the Catholic school board. Along with the typical music and sports extracurriculars, the academic program there was rigorous, and my science classes were filled with smart women and teachers who expected us to grasp it all—and succeed. There was never any talk of "women aren't good in math" or crap like that. It never occurred to me that my experience was somewhat anomalous—I was in for a shock later on.

I also met the best friends a girl could ask for in high school: Alwynn, Brandee, Charlene, and Heidi. (At some point, someone affectionately called us "The Broad Squad," and it stuck.) These women taught me so much: how to be confident, vulnerable, and comfortable in my own skin. They also taught me the filthiest swear words you could possibly imagine, and we went through the typical growing pains, trials, joys, and heartbreaks of high school (and life afterward) together. As members of the school band executive committee, we spent many a day in the "band office," which in a previous life was a walk-in freezer tucked in a corner of the basement next to the music room.

We'd sit on the grimy couch and broken chairs, our kilt-and-cardigan uniforms in various levels of disarray, listening to our favorite musicians—I think at the time it was Sarah McLachlan and Salt-N-Pepa—and talk about boys and play card games like "Big Hoss" between classes—and sometimes, instead of going to classes. But somehow I managed to keep my grades up.

Although my science and math skills convinced my parents that I was going to be a medical doctor, all thoughts of a career in medicine were quickly shattered on a particularly gruesome day of tenth-grade biology, when we were tasked with dissecting the scariest-looking cricket on the planet. There I was in my goggles, lab coat, and latex gloves ready to meet Jiminy when, instead, my teacher slapped down an aluminum tray filled with the rankest formaldehyde-soaked, foot-long, segmented slimy, three-mouthed (OK, probably not) alien. If memes had existed back then, I'd have been the octopus scuttling away with the "nope, nope, nope" phrase written underneath. Every part of my being was convinced this thing was going to jump in my face. The sensation of sliding a scalpel into a squishy body filled with putrid fluids is not as exhilarating as all the television ER dramas make you think. My lab partner ended up taking the lead on the dissection. She also ended up doing the pig dissection later in the course. Thanks, Heidi.

So, how did I end up becoming a planetary scientist instead of a medical doctor? I wasn't the type who stared at the stars as a child and knew I wanted to study planets. I was more like a ball rolling downhill, catching speed while dodging obstacles and varied topography, with a hole dug somewhere along the way to finally stay at rest (for the physicists: think local energy minimum). That was me—bouncing around, loosely guided by my interests, and finally landing in a comfortable crater, filled with my converging interests instead of rich minerals.

Three people strongly influenced my life and career choices during those formative high school-university years. The first was fictional: the esteemed starship figurehead Captain Jean-Luc Picard. I had a standard after-school routine in my teen years: throw the plaid kilt on the floor, settle in my favored spot on the living room carpet in my tights and cardigan, turn on the TV, and start my homework as reruns of *Star Trek: The Next Generation* played. Admittedly, I was initially guided less by fandom and more by it being the only thing on TV that wasn't sports or news.

In the twenty-fourth-century world of *Star Trek,* filled with cooperation, exploration, and scientific inquiry, I found kindred spirits—a future where humans seemed to make much more sensible decisions on what was important in life alongside their alien allies. That spirit of charting unknown worlds hooked me on space and made me want to be part of that greater purpose.

The second person who influenced my life and work was my high school music teacher, Mr. van Raadshooven ("Mr. Van" to us, as he also drove a van—we loved this congruency). I spent most of my time in the music room, practicing piano, alto sax, and oboe, with notes from Mr. Van excusing me from class. He went above and beyond for students and inspired my love of music; he would also later inspire my love of teaching. (I briefly considered a career in music but didn't pursue it—to the relief of my mother, who spent part of her twenties touring parts of Germany with her rock band.)

I met my third mentor in 1995 on my first day at the University of Toronto when I entered the auditorium for my Foundations of Physics class. I was one of about five women to a hundred men in an ice-cold auditorium with rock-hard, salmon-colored chairs. Considering my previous schooling, my first thought was, "Why are all these guys here?" (Ten years later, I'd teach this very class in this very hall. Unfortunately, the gender balance hadn't gotten much better.) Our professor was Jerry Mitrovica, who had received previous teaching reviews like "Best prof ever!" and "I'll be naming my firstborn after Jerry." And despite the brutal nature of intro physics, everyone who had taken the class the prior year said they'd take it again.

I hoped he'd live up to the hype. He definitely did. Jerry became an amazing influence. There were the fundamentals of teaching, like help with problem sets and skilled guidance answering questions during office hours. But there was also the human element, the Jerry who would regale us with stories of his life, filled with hilarious flaws and foibles that brought him down to earth. His advocacy for me extended into helping me connect with researchers who were focused on the sorts of topics that interested me, first cosmology and then planetary science. These opportunities, in turn, ultimately led me to graduate school at Harvard. Now we work in similar fields and have coauthored papers; he remains a mentor and dear friend.

Jerry still likes to tell this story: Around the time of my sophomore year, I thought I wanted to work on cosmology, a field involving the origins of the universe, and was looking for research internships in string theory or maybe black holes (or other hot topics from that time). I asked him for a reference letter for a prestigious summer research opportunity at the Canadian Institute for Theoretical Astrophysics (CITA). He sized me up: pink hair and a nose ring—not exactly the model of most physicists at the time. Despite an A- in his course, I wasn't one of the stars of the class, either. He prepared to tell me I'd be a long shot and should consider some other options, trying to let me down easy. But then he looked at my transcripts, with stellar marks in all the right courses like calculus, linear algebra, and astrophysics.

At that moment, he says, he stopped himself, having realized that his initial perception of me had nothing to do with what I was capable of, and moved past his first impressions. He wrote a strong reference letter for me, and though I didn't get the position, his vote of confidence meant a lot. It also instilled an important lesson in me: We all have preconceived notions and biases, but when confronted with data about the real world—and the universe beyond it—we need to alter our worldview.

It was a good thing that I didn't get the position, ultimately. In my junior year, I'd taken a course in cosmology, and the fact that these objects were so far away—almost intangibly so—made me realize it wasn't quite what I wanted. Planets seemed like the better fit, at a time when astronomers were discovering the first worlds outside our solar system (1992), NASA was studying the Jupiter system with the Galileo orbiter (1995), and the United States had landed the first rover on Mars (1997). When I shared my new ambition with Jerry, he guided me toward two researchers at Harvard—one working on seismology studies of tsunamis on Earth, the other studying the magnetic fields of Uranus and Neptune. I knew nothing about the topics, but ultimately chose the latter for a summer undergraduate research position. This was where I really began to learn about the complexities and beauty of the planets' interiors. I ended up loving the project so much that I went back to Harvard for graduate school to continue my studies.

After Harvard came a postdoc at the Massachusetts Institute of Technology, where I investigated Mars's and Mercury's interiors and how they generate magnetic fields. Then I held a professorship back at the University of Toronto for about 12 years. I finally arrived at Johns Hopkins University as a Bloomberg Distinguished Professor in 2017.

During this time, I've been fascinated with comparing the inner workings of planets: Why are some planets so similar in their interiors? Why do some have such odd behavior? Why is one planet (Earth) teaming with life, while others appear to be barren? I've published research on the interior of almost every single planet in our solar system (Venus: I'm still coming for you!) and even some research on planets outside of our solar system. I'm still awed by the fact that every new planetary mission we send to a neighboring world brings us data that challenges our paradigms while bringing us closer to understanding how planets work.

A fundamental tenet of science is that with better data we'll have better theories to explain the phenomena we study. As someone who studies planets, I'm reminded of this often. But with planets, some data sets are easier to come by than others. The hardest place to study is that which is hidden from us: the planet's interior. When we send missions to other planets and moons, we can image the surface (unless the atmosphere is a real pain like with Venus and Saturn's moon Titan). And we can use the interaction of light with matter to study what's going on in a planet's atmosphere. But a planet's interior? That's a trickier beast. But there are ways, and they'll be detailed in this book.

Arguably, a planet's interior is more important than the surface in determining a world's fitness for life or ability to withstand the pressures put on it by its home star. Nine times out of ten, the interior has major implications for everything going on at the surface. To understand a planet, we thus have to interrogate what's going on below its surface, a frustrating endeavor considering that we can't directly see the interior and can only drill down fewer than ten miles into Earth.

There's an entirely different existence right beneath our feet here on Earth. In these invisible depths, materials get crushed in extreme pressures, creating new phases of materials we've never experienced on Earth's surface. Some planets' interiors have diamond icebergs, iron snow, and helium rain. The interior of a planet is just going on living its best life, not caring if we are in on its secrets, but we need to craft ways to understand what's hidden. Then we can begin to reveal the missing clues to answer some of the most fundamental questions in planetary science: How do planets form, how do they evolve, and how can planets become hospitable to life?

In this book, I hope to bring some of these secrets to light. I'll consider it a success if someday after reading it, you'll be walking down the street, look down at your feet, and think to yourself, "Wait a second. The world doesn't end there—there's even cooler stuff happening right now a thousand miles below me." So join me in this journey to the center not just of Earth but of Venus, Mars, Jupiter, and so many other fascinating and mysterious destinations, too.

What's Hidden Inside Planets?

CHAPTER 1

Gazing Inward

IT WAS 2 A.M. ON A SUMMER NIGHT. I had just arrived home from a night out with friends when I stepped out of my car and my jaw dropped. The dark sky was lit up with vibrant green flashes, and I stared, mesmerized, at the elegant motions. This was my first time seeing the "northern lights" of the aurora borealis in person. Although my hometown, Sudbury, is considered part of northern Ontario in Canada, it isn't so far north as to frequently host light shows like this.

Auroras are some of the most beautiful cosmic spectacles visible to the naked eye from the surface of Earth. They provide an amazing illustration of the interplay between our Sun, 94 million miles away from us, and Earth's molten iron core, 1,800 miles beneath our feet. It's in the core that Earth's magnetic field is generated, and this is crucial to the creation of auroras, which hover anywhere from 60 to 600 miles above our surface in the ionosphere.

About 25 years have passed since that night, and I've spent a good chunk of that time studying the generation of Earth's (and other planets') magnetic fields. I find it fascinating that most of the conditions we experience that make life comfortable here on Earth's surface, like our magnetic shield, are actually the result of processes occurring deep in our planet, hidden from view.

Deep beneath our feet, too, nutrients and volatiles (chemicals that readily evaporate, like water and carbon dioxide) cycle and are ultimately responsible for the creation of our atmosphere and oceans. They also guide the regulation of our climate and surface temperature through the carbon cycle. None of those would be possible without the machinations of Earth's deep interior.

But there is a dark side, too. Some of the most dangerous natural hazards on Earth—earthquakes, volcanoes, tsunamis—are also the by-product of deep-Earth processes. This gives us all the more reason to try to understand what goes on deep beneath our feet.

And as the auroras demonstrate, the insides of Earth also interact with the space surrounding Earth. When it comes to auroras, it's an interplay between the Earth and the Sun. The Sun spews out a torrential wind of fast-moving ions and electrons, known as the solar wind. Those solar wind particles head to Earth at about a million miles per hour and would slam right into the planet's surface if it weren't for Earth's magnetic field. But Earth's core generates a magnetic field able to fend off those dangerous high-energy particles.

If not for our core, the onslaught of solar wind particles would devastate life as we know it on Earth. These high-energy particles can create cell damage that causes cancer, erode our atmosphere, and induce power surges and blackouts by blowing transmission lines.

I like to imagine Earth's magnetic field as an invisible shield that some superhero has put up to stop the onslaught of solar wind. In reality, that superhero is a physics process known as a dynamo—a conversion of motion into electromagnetic energy through the creation of electrical currents. And it's the same dynamo that can explain how your home generator works, or why pedaling can provide energy for your bike light.

Dynamo bike lights are a great demonstration of how you can take energy in one form and convert it to another. As you pedal, the bike light mechanism (the "dynamo") creates electric currents that flow in a wire. So the kinetic energy—the energy associated with motion—of the pedaling creates electromagnetic energy that ultimately lets the bike light cut through the darkness.

The basic idea is the same in Earth's core, although without the wires. So what process is doing the equivalent of "pedaling" in Earth's core to create kinetic energy from motion? I've seen many websites and even some scientists claim that Earth's dynamo is created by its spinning. But that's not the case. While it is absolutely true that the Earth is spinning, that's not the motion that a dynamo needs to ultimately create magnetic fields.

Instead, buoyancy drives convective motions that create our protective magnetic field.

Having grown up in a family that loves food, I'm drawn to cooking analogies when talking about the Earth and how it works. The convection from buoyancy forces in Earth's core is an easy one: Let's make a pot of soup. Take your ingredients and put them in a pot, then heat them on a burner. As the flame heats the bottom of the pot, you set up a difference in temperature between the bottom and top of the soup. Even before the soup comes to a boil, you can see motions occurring in the broth as the hotter liquid close to the bottom of the pot near the burner moves upward.

The key thing here is that hotter materials expand. This is true for the air in your tires, the water in the ocean, and even the lids on your jars that are stubbornly fused together until you run the lid under some hot water. It's also the reason hot air rises. The hotter broth at the bottom of the pot is therefore a bit expanded, making it slightly less dense than the soup above it that isn't as close to the burner. Its reduced density makes it weigh less than an equivalent volume of surrounding soup, and so it becomes buoyant and floats upward. Once it reaches the top, it can cool down a little bit, losing its buoyancy, and the cycle repeats, resulting in the observable rolling motion that we call convection.

The churning molten iron in Earth's core is like the soup. The inside of Earth (and all the planets) is hotter than the outside. That's because planets are able to trap heat inside of them as they form. Planets also create heat by the natural decay of their radioactive elements. As our planet cools to space, the liquid in Earth's core convects. There's no burner at the center of the Earth, as there was under the pot, but the key thing is that the temperature at the bottom of the liquid core is hotter than at the top, so the iron at the bottom expands a bit, becomes less dense, and can float upward, creating convective motions. Because the Earth's core is made of liquid iron, unlike your soup (hopefully), the motion of the liquid iron can induce electric currents and ultimately create Earth's dynamic magnetic field.

Although humans can't see or feel Earth's magnetic field, some animals can. Magnetoreception is used by animals such as migratory birds and fish like sharks and salmon to help with orientation and navigation. There is even evidence (although controversial) that cows and deer align their bodies with Earth's magnetic field while they're grazing or resting.[1]

And even though humans don't directly use Earth's magnetic field for navigation, we do use it through tools and technology. For example, some commercial satellites, like the Iridium constellation of telecommunication satellites, use the field to help determine their orientation. And planes and ships rely on it for navigation. Most of our navigation apps include information from the World Magnetic Model, which includes an accurate description of the Earth's magnetic field.

Earth's Superpower: Magnificent Magnetism

Earth's roiling outer core of molten iron transmits electromagnetic energy through thousands of miles to the surface and beyond, ultimately forming an invisible shield—a kind of geomagnetic superhero cape, blown around by solar winds. This magnetosphere surrounding Earth, made possible by a dynamo process in its core (for more information, see page 9), protects our atmosphere from erosion, and repels energies such as charged solar plasma radiation and cosmic rays that threaten life and industry. But superpowers have their kryptonite; occasionally, intense solar wind variations can pierce the shield, leading to disruptions in electrical grids, navigation, satellites, and telecommunications. But most of the time, the shield is a guardian with two standout benefits: its application to navigation, and the auroras.

The Aurora Polaris: The flipside of solar disruptions to the magnetosphere is the wonder of these swirling, twisting light shows that travel along magnetic field lines into Earth's upper atmosphere, the ionosphere. These auroras are powered by charged electrons and protons colliding with atmospheric molecules that release tiny light flashes visible as the greens (from oxygen) and blues and purples (from nitrogen) at certain latitudes. The northern lights are known as the aurora borealis, while those near the South Pole are called the aurora australis.

While we think of these effects as visual displays, people around the world have also described a sound when they're seen, akin to the snapping of fingers, rustling grass, or the crackling of radio static. Nordic Sami people call the lights *guovssahas*—the light you can hear. While folk accounts of these sounds over the centuries have been met with skepticism, in 2012, auroral acoustics researcher Unto Laine of Finland's Aalto University recorded them about 230 feet above ground level. In 2020, some of those recorded sounds were incorporated into a work for string trio by Irish composer Sam Perkin that premiered at Festivalta, a Norwegian chamber music performance.

It's no surprise that such displays inspire storytellers worldwide; the Earth and its mysteries, and our desire to understand them, have united us for millennia. The ancient Chinese saw the light as the fiery breath of battling dragons, while Australian aboriginal people considered them omens of fire and bloodshed. The Inuit of Nunivak Island in the Bering Sea offered one of the more colorful interpretations: that the phenomenon was a huddle of walruses playing a ballgame with a skull.

Navigation: Birds, sea turtles, and sharks, along with many other creatures, are believed to use "magnetoreception"—in addition to celestial body positions, scents, and landmarks—for orientation, homing, and navigation during migrations. This concept of internal compasses is a growing area of research, but no one knows for sure—yet—what the mechanisms are that keep creatures on course for sometimes thousands of miles. Scientists at Oxford University are studying cryptochromes, a magnetically sensitive chemical in birds' retinas, that might hold a clue.

What's known for certain is that one species of mammal—humans—has used the magnetic field to orient and navigate following the invention and refinement of the compass, believed to have been introduced in a more primitive form by the Chinese around 200 BCE during the Han Dynasty. The initial tool was created using a magnetic mineral called magnetite for configuring spaces according to the precepts of feng shui. It would be more than a thousand years before it was refined for navigation, eventually leading, along with shipbuilding and other innovations, to the age of discovery.

Little did those sailors know that what made their voyages around the globe possible through the accuracy of those compasses was a scorching ball of iron thousands of miles below their hulls.

FURTHER READING

"Aurorae Australis." *Australian Indigenous Astronomy.* Accessed March 22, 2023. http://www.aboriginalastronomy.com.au/content/topics/aurorae/.

Dobrijevic, Daisy. "Earth's Magnetic Field: Explained." *Space.com,* July 6, 2022. https://www.space.com/earths-magnetic-field-explained.

Metcalfe, Tom. "How Scientists Searched for the Elusive Sounds of the Northern Lights." *NBC News,* October 11, 2021. https://www.nbcnews.com/science/science-news/scientists-searched-elusive-sounds-northern-lights-rcna2840.

Warner, Kylie. "How Some Animals Use the Earth's Magnetic Field to Navigate." *The Economist,* September 25, 2018. https://www.economist.com/the-economist-explains/2018/09/25/how-some-animals-use-the-earths-magnetic-field-to-navigate.

But how does our planet's dynamo create the magnetic shield, extending a distance of about 40,000 miles from Earth in the direction of the Sun, that protects life on Earth? There are two important factors here. First, the Earth's magnetic field is predominantly *dipolar*. This means it has two magnetic poles, at the far northern and southern reaches of the planet. Magnetic fields have directions, and for a dipolar field, the field lines emanate from near the northern magnetic pole and travel in an arc-like shape until they reach the southern magnetic pole. At present, the Earth's magnetic south pole is near the geographic north pole and vice versa.

The second important factor is that charged particles, like the electrons and ions that make up the solar wind, must circle around magnetic field lines. It's as if they have tiny strings that they hook onto magnetic fields and get spun around at the end of them. The hooking point can move along the magnetic field line, but the particles keep moving in circles around the field line, resulting in a spiraling motion toward one of the poles.

So now we combine these two important factors—the dipolar field and the spiraling of charged particles—to explain how Earth's magnetic field shields us from the solar wind. As the solar wind approaches Earth at a speed of about a million miles per hour, rather than slamming into the planet, the charged particles get trapped away from Earth because they spiral along the magnetic field. That forces them to head toward

The Magnetosphere's Mechanisms

High-energy particles in the solar wind, mainly electrons and protons, are deflected from Earth by the magnetic field created in the Earth's core, like the deflector shields in *Star Trek* protect a Federation ship from enemy torpedoes. High-energy particles become trapped along magnetic field lines, spiraling along them until they reach polar regions where they collide with air molecules such as nitrogen and oxygen, creating the auroras.

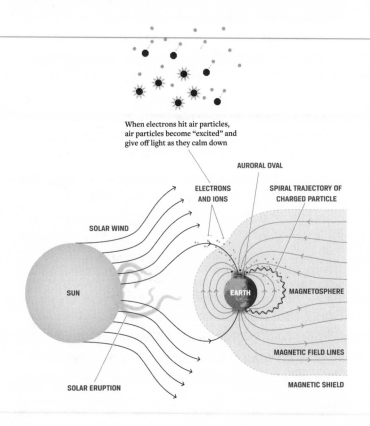

When electrons hit air particles, air particles become "excited" and give off light as they calm down

AURORAL OVAL

ELECTRONS AND IONS

SPIRAL TRAJECTORY OF CHARGED PARTICLE

SOLAR WIND

SUN

EARTH

MAGNETOSPHERE

MAGNETIC FIELD LINES

SOLAR ERUPTION

MAGNETIC SHIELD

the north or south pole, encircling a magnetic field line. At the "mirror point," the magnetic field near the pole gets so strong that the charged particle actually turns around and heads to the opposite pole.

All the magic that we see in creating the auroras happens when those field lines approach the poles. The poles are the only place where the field lines head toward the surface of the planet. But before they reach the surface, they encounter Earth's atmosphere. Some of the solar wind particles penetrate deep enough into Earth's atmosphere near the poles that they can collide with unsuspecting atmospheric gas particles, like atomic oxygen. That collision transfers energy from the solar wind particle to the atmosphere particle, and the atmosphere particle gets knocked into a higher energy state. After such a rude encounter (I picture the exasperated oxygen atom waving a fan and proclaiming, "well I never!"), the oxygen atom eventually settles back down to its regular energy level by emitting a photon of light. For oxygen at low altitudes, those photons happen to produce green light, the most commonly observed aurora. Other colors, like red, pink, blue, and purple, can be produced at different altitudes in the atmosphere, from collisions with different gases, such as nitrogen, and from combining colors.

Although the auroras typically happen closer to the poles, sometimes they're produced at lower latitudes because Earth's magnetic shield gets deformed if it can't stave off an overactive solar wind—one caused, for example, by a coronal mass ejection or solar flare erupting on the surface of the Sun. Such an event occurred that night in Sudbury, causing a geomagnetic storm and allowing me to witness the flashes of green sheets in the sky.

The largest geomagnetic storm on record occurred in 1859 and is known as the Carrington event, named for Richard Christopher Carrington, who was one of the first astronomers to identify solar flares. Auroras were witnessed as far equatorward as Mexico, the Caribbean, and Hawaii. The auroras were so bright in parts of the United States that gold miners in the Rocky Mountains apparently woke up and cooked breakfast at 1:00 a.m.[2]

If such an event were to happen in our technologically dependent world today, it would be devastating. There would be electrical blackouts globally, radio communications would be shut down, and satellites in orbit around Earth would be irreparably damaged.

You might be wondering how often these Carrington-like events happen. Their cause, coronal mass ejections (CMEs), happen quite often and are tied to the sunspot cycle of 11 years. During solar minima, when there are few sunspots, they happen about once a week; during solar maxima, they can occur daily. But not all CMEs are created equal, and it's hard to predict when one big enough to cause a Carrington event will occur. A similarly sized CME occurred in 2012, but fortunately it didn't erupt in our direction, so it missed hitting Earth by nine days.[3]

If we'd been in a slightly different location in our orbit, we would likely still be recovering from the devastation of the event today.

Admittedly, when I watched the auroras that night, I didn't give a second thought to the role Earth's core plays in generating auroras; I was too captivated by the beauty of the display. But I suppose that's the point of this book: I hope to convince you that while you spend time awestruck by some of the universe's and nature's wonders, by looking up and looking around, that it's important to sometimes "look" inside, too, beneath the soil, sand, and water of our home planet.

DRILLING INTO EARTH

So now perhaps you're looking down at your feet and the ground below them. What's under that surface? The Japanese poet Kobayashi Issa (1763–1828) wrote my favorite haiku on the subject, which in English translation reads:

> In this world
> We walk on the roof of hell,
> gazing at flowers.

The concept of a world beneath our feet was and is prevalent in many cultures. As a big fan of Earth's interior, I'd argue it's grossly unfair that most descriptions of hell place it underground, while heaven is usually lofted into the sky. Of course, it's true that if one were to be transported deep underground, the extreme heat and suffocating pressure would truly be unbearable, but being transported to a high altitude with extreme cold and very low air pressure would be just as uncomfortable.

Our experiences as humans make it challenging to fully grasp the conditions that occur inside the planet. There's just nothing like it in our everyday experience. It's even challenging for scientists to recreate the environment of the Earth's interior in state-of-the-art laboratories, although we've had some success with giant lasers at high-energy particle colliders. In order to understand what we might expect to happen deep inside our home, let's start with some basic principles.

Earth is, roughly speaking, a round ball. We'll talk about how deviations from a perfect sphere actually tell us a lot about the inside of Earth in later chapters, but for now, approximating the Earth as a sphere can be useful. It has a radius of about 4,000 miles and can be divided into spherical shells, or layers, with different properties. How you divide it (for example, by composition, or mechanical properties) results in different names for the layers, but we can't just dissect the Earth to examine these layers. Instead, scientists have found clever ways to probe the insides of the Earth using tools similar to what doctors might use to study the insides of the human body.

Earth's Layers

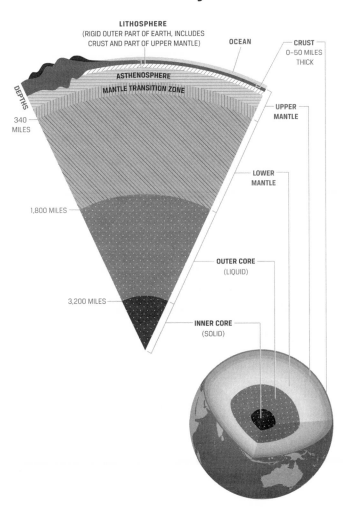

LITHOSPHERE
(RIGID OUTER PART OF EARTH, INCLUDES
CRUST AND PART OF UPPER MANTLE)

OCEAN

CRUST
0–50 MILES
THICK

ASTHENOSPHERE

MANTLE TRANSITION ZONE

DEPTHS

340 MILES

UPPER MANTLE

LOWER MANTLE

1,800 MILES

OUTER CORE
(LIQUID)

3,200 MILES

INNER CORE
(SOLID)

Our first instinct on how to determine what's inside is to dig a hole into the planet and sample it (hopefully this isn't your doctor's first instinct in trying to determine what ails you). For Earth, and all planets, this is easier said than done.

HOBBLED JOURNEYS TO THE CENTER

We humans are natural explorers. We've sent spacecraft to visit every planet in our solar system, as well as a handful of asteroids, comets, and moons. We've even touched the atmosphere of the Sun with the Parker Solar Probe. The record for the farthest distance a spacecraft has traveled from Earth is currently held by *Voyager 1*; at the moment of writing it is 14.8 billion miles away, probing the boundary between our solar system and interstellar space. This boundary is created by the Sun's magnetic field and shields our solar system much the same way that Earth's magnetic field shields Earth. With such astronomically large numbers for distances governing our technological capabilities for exploring, it's somewhat surprising that the deepest we've ever been able to reach inside the Earth for exploration is less than eight miles. This isn't for lack of trying or curiosity. Indeed, great sci-fi and fantasy stories have been generated on the topic, such as the classic *Journey to the Center of the Earth* by Jules Verne and my favorite, the 2003 movie *The Core*.

As a child, I spent hours in my backyard sandbox digging with a bright yellow shovel wondering how deep I could get.

Unfortunately, I was quickly thwarted by the timbers that made up the bottom casing. The few opportunities availed to me while on family vacation to dig in the "real" sand on a beach were squandered by a combination of sheer exhaustion after the effort of reaching a couple of feet down in the sand and the many other fascinations of an ocean beach, such as searching for seashells or splashing in the waves. This last experience is most likely responsible for nucleating my love of fluid dynamics, the study of which is fundamental to understanding the physics of the Earth and planets.

Most people are familiar with the "space race" between the United States and the Soviet Union that ultimately led to the Americans landing astronauts on the Moon in 1969. Perhaps fewer people know that there was also a "mantle race" during the Cold War: Which of the superpowers could drill through the Earth's crust and reach the mantle first?[4]

The early contenders were, not surprisingly, the United States and the Soviet Union. The motivations: the search for new oil and mineral resources, the scientific exploration of the rocks below the surface, and a chance to prove the superior scientific capabilities of the respective nations. Holes were drilled in a variety of locations for different purposes.

The United States aimed to reach the boundary between Earth's crust and mantle, called the Mohorovičić discontinuity, or "Moho" for short, with Project Mohole, which began in the late 1950s. The project was initiated by a group with what might be the best collaborative name ever: the American Miscellaneous Society (AMSOC). The society itself didn't do much except for comprising leading scientists of the time who, as far as I can tell, mainly wanted to rebel against the rules and strictures of typical professional societies. But members of the society eventually took on the challenge of Project Mohole, aiming to pitch it as a grand challenge in Earth science, similar to the grand challenge in space science of reaching the Moon. The American Miscellaneous Society, which led the effort, was smart about the approach, choosing a spot in the ocean where the crust is relatively thin near Guadalupe Island in Mexico. The first phase of the project in the early 1960s was quite successful in demonstrating new ocean-based drilling technologies and reaching a maximum depth of about 600 feet below the ocean bottom. There was enormous excitement surrounding the achievements of phase I— John Steinbeck, a lover of oceanography, even accompanied the ocean voyage, documenting the efforts in *Life* magazine: "This is the opening move in a long-term plan of exploration of the unknown two-thirds of our planet that lies under the sea. We know less about this area than we do about the moon." He ends the piece stating: "They have opened the way to the exploration and eventual development of the greatest part of our home planet. And I feel a joy like a bright light at having been there to see."[5]

Sadly, the project was a victim of bad management, with infighting and cost overruns that resulted in its shutdown. In a commentary piece for *Science* magazine titled "Mohole: The Project That Went Awry," D. S. Greenberg refers to it as "a classic case of how not to run a big research program."[6] I highly recommend this article (which includes such gems as "wine breakfast," and "The ocean's bottom is at least as important to us as the moon's behind") if you want a more detailed account of the whole affair.

In 1974, an oil wellbore reached almost six miles deep after a year and a half of drilling in Washita County, Oklahoma. This held the record for the deepest borehole until 1979, when the Soviet Union's Kola Superdeep Borehole in northwestern Russia surpassed it. Drilling continued there, eventually reaching a depth of 7.6 miles in 1989. To date, the Kola borehole still holds the record for the deepest artificial hole in Earth.

Because the location of Kola Superdeep* is the thick Baltic Shield, there was a lot more crust to get through here before reaching the mantle. That was intentional, as scientific studies of the old, deep crust were a major driver of the project rather than, say, beating the United States to reaching the mantle. Although drilling continued past the record set in 1989, it required off-shoot holes due to breakdowns in equipment and high temperatures in the original hole. No more progress was made to reach further depths, and the drilling stopped in 1995.

Kola Superdeep is also the name of a 2020 Russian horror film that I didn't know existed but is now on my watchlist.

Other nations began scientific drilling programs, like the Kontinentales Tiefbohrprogramm der Bundesrepublik Deutschland, or KTB, in Bavaria, Germany, which operated from 1987 to 1995. Since 1996, nations have worked together as part of the International Continental Scientific Drilling Program (ICDP) to study the Earth's crust and the related resources (like precious metals and minerals) and hazards (like earthquakes and volcanoes) associated with it. The International Ocean Discovery Program, a collaboration between science agencies in the United States, Japan, the European Union, China, Australia, New Zealand, and India, pursues scientific drilling projects from ships and platforms in ocean areas. These programs use scientific results from drilling projects to answer questions related to climate change, tectonics, earthquake hazards, life in the deep biosphere, and alternative energy resources.

No one has yet drilled through the Earth's crust and reached the mantle, so the mantle race has yet to be won. I don't see as much fervor for the effort now as there was in the space-race era. Although there are a few places on Earth where the mantle reached the surface because of how Earth's plates have shifted around (for more on this phenomenon, see page 108), that's not "pristine" mantle in its natural habitat. But studying the crust is also important; there is much still to learn from the outermost layer of Earth's interior, and good science is happening in these pursuits related to our understanding of topics such as plate tectonics and Earth's history.

THE CHALLENGE OF EXPLORING DEEP EARTH

Why is it so hard to dig, or drill, or tunnel, deep into the Earth? The answer is ultimately related to the fact that the deeper one goes into the Earth, the higher the pressure and temperature. It's a really extreme environment.

You can get an inkling of the conditions if you consider the issues confronted by scuba divers. Our bodies have evolved to maintain a very important balance: the pressure surrounding our body and the pressure inside. The latter is due to all the atoms we're made up of randomly vibrating in the fancy molecular or crystal structures in which they exist. Also the cavities, such as our lungs, expand and contract when filled with gases or fluids. The pressure outside our bodies is due to the atoms around us and their random motions running into us. Take the atmosphere, for example. The atoms and molecules in our atmosphere are constantly in motion and colliding with our bodies. That's what gives us the "feel" of atmospheric pressure. Atmospheric pressure is usually measured in the "bar" units, and at Earth's surface, the average atmospheric pressure is 1 bar, equivalent to 14.5 pounds of force per square inch.

What creates that pressure? When I first learned the answer, it simultaneously made perfect sense and seemed completely ridiculous. What we feel is due to the weight of all of the atmosphere in a column that goes from right above you to the top of the atmosphere, which is about 500 miles above you. I should warn you that there are many ways to define the "top" of the atmosphere. Here I've chosen to use the base of the exosphere, which marks the altitude at which atmospheric particles are so far apart that collisions between them are extremely rare and have very little effect on the pressure at Earth's surface. The pressure you feel at Earth's surface is therefore due to the fact that you are essentially carrying a column of air, 500 miles high, on top of your head at all times.

~~So why aren't we crushed by this? Because we evolved in~~ these conditions, and our entire system has evolved to withstand it. But of course, there are tolerances, we can climb mountains (where the column of air above our head is a little shorter so the pressure is lower) without exploding outwards and we can swim underwater (where you have to then add the weight of the column of water above you to the air above that in order to figure out what pressure you feel) without being crushed.

But we can't swim too deep underwater without added safety measures to deal with the increasing pressure. About every 30 feet underwater corresponds to an added atmosphere of pressure. The deeper you dive, the more air pressure you need to breathe in order for your lungs not to collapse. A free diver without any equipment, like the pearl divers of Japan, can go as deep as 100 feet (although the record for the deepest free dive is 700 feet) while scuba divers with air tanks can reach maximum depths of around 1,000 feet.

Now think about what would happen if we could "dive" into the rock of Earth. Rocks weigh about three times as much as water near Earth's surface, so the pressure increases much more quickly with depth as we go into the Earth. Perhaps you're thinking: let's be reasonable here. We aren't going to just swim into the Earth; we can build tunnels or stable holes where the only thing above us is a bit more atmosphere. It sounds like a great idea, but there are two problems, both related to how pressure increases with depth.

The first problem is that increased pressure means increased temperature, and the temperature increases inside the Earth at a rate of about 70 degrees Fahrenheit for every mile you descend. That's a huge burden on human miners, equipment, and air-conditioning systems in deep mines. Currently, the deepest mine is the Mponeng gold mine in South Africa, which operates at a depth of about two miles. Given that the center of the Earth is about 4,000 miles down, that means that these mines barely scratch the surface when it comes to exploring the interior of the Earth.

The second problem is structural integrity. Tunnels and boreholes work because materials like rocks, or metals, or whatever you think of that's "strong," have cohesion: their structure is pretty stable against being deformed and compromised. If you create a tunnel under a bunch of rock, the cohesive strength of the tunnel's roof has to be strong enough to deal with the weight of all the rock lying above it. Rocks are strong but not infinitely so. Go deep enough and the weight of the overlying rock wins, and the tunnel will collapse. The same is true if you try to build a fancy capsule out of a fanciful material, like the fictitious "unobtainium" in the movie *The Core*.

JOURNEY TO THE CENTER OF THE EARTH

But let's do a little thought experiment. Let's say we could line our tunnel or build a drilling vessel in which to travel as we descend through the layers out of the strongest material we have on Earth: graphene. Graphene is a form of carbon in which the atoms form in a single layer with a honeycomb-like structure. It has been measured to have a tensile strength of around 1 million bars (that's 1 million times Earth's atmospheric pressure and five times stronger than diamond), and over the past decade, its strength, lightness, and conductive properties have resulted in fervent research to determine its potential for use in everything from better transistors to solar cells to condoms.

Even if humans became savvy enough to build a ship out of graphene to explore the interior of Earth, 1 million bars only get us about halfway to the center of the Earth! Anything deeper than that, and our graphene ship would crush everything within it and would also probably transition to a different phase of carbon.

Let's expand our experiment so we don't have to rely on the limits of our current technology. Assume we find some material from which we could make a vessel to take us on a trip to the center of the Earth. Our journey begins in the outermost layer of Earth: the crust. Due to proximity, this is the layer we have the most experience with and know the most about. If we choose our descension point to be on a continent, then the crust there is on average, only about 20 miles thick—that's about 0.5% of the Earth's total thickness. If we choose a point at the bottom of the ocean to start, then the crust is only a few miles thick. To once again reference *The Core*, in terms of thickness, the crust of the Earth is comparable to the skin of a peach.

Earth's crust forms when hot rocks from deeper inside Earth reach the surface and cool. In the oceans, this occurs at places where Earth's tectonic plates are being ripped apart: mid-ocean ridges. On continents, crust forms when blobs of hot rock rise up and either spew onto the surface from volcanoes as lava flows or emplace themselves as magma blobs under the surface—this is like a volcano that erupts underground.

The chemistry and mineralogy of these rocks is fascinating and can take a lifetime to explore—just ask a geologist. The crust is also the youngest part of Earth. Most of Earth's surface is much younger than the planet itself. It turns out that this is a somewhat unusual feature of Earth among the planets. Most planetary surfaces are quite old, although Venus is another exception. Much of the crust of the Moon and Mars, for example, is around 4 billion years old. On Earth, the oldest crust under the oceans is only about 200 million years old, whereas crust on the continents has an average age of about 2 billion years; there is very little crust more than 3 billion years old.[7]

The absence of old crust is actually quite frustrating if you're interested in studying the early history of Earth. There's just not that much rock around from that time period. We've probably learned more about the early history of Earth by studying the Moon, Mars, and Mercury, all of which have lots of old crust, than we have from studying Earth itself. The crust is also the least dense, or lightest, part of the Earth's interior, which is why it has floated to the top of the Earth. Once again, it's all about buoyancy, just like in the Earth's core.

Let's get below the crust as we continue our descent into Earth. As seen in the figure on page 39, the next layer we reach is called the mantle, and it makes up about half the radius of Earth. It's a little bit misleading to think of this as one giant homogenous layer. The mantle changes quite a bit as you descend into the planet. That's because the pressure and temperature increase significantly with depth, and that affects the stability of minerals in the rocks that make the mantle.

Just as the element carbon can transform from the graphite in your pencil to diamond at higher pressure to graphene if produced in a lab by rearranging how the atoms are bonded in its crystal structure, the minerals that make up the rocks in the mantle can undergo structural transformations, known as phase changes. These can affect the properties of the rocks, like their density and how easily they deform.

Near the top of the mantle, the pressures are around 10,000 bars (that's 10,000 times Earth's surface pressure), and temperatures are a ghastly 1,800 degrees Fahrenheit. By the time you get to the bottom of the mantle, which is about halfway down to the center of the Earth, your pressure is over a million bars, and your thermometer would read about 7,000 degrees Fahrenheit.

Below that, we reach my favorite layer of the Earth, the core, which makes up the innermost 2,000 miles or so. The core is quite different in composition from the mantle and crust. It's mainly an iron-nickel alloy. We know this not because we have samples from Earth's core, but because we can use information from studies of gravity, seismology, and meteorites to determine the density of the core (I discuss these methods in later chapters).

Notice that I said *mainly*. It turns out that Earth's core is a bit lighter than it would be if it were pure iron-nickel. There are, in fact, some lighter elements mixed in. Scientists debate the exact amount and elements that are stuck down there in the core, with everything from hydrogen to sulfur, oxygen, and carbon being suggested as likely candidates. Whatever they are, these light elements make up about 10% of the core's weight. These light elements are crucial to core convection and hence to the generation of Earth's magnetic field.

As we travel into the core, we start in the liquid part, which makes up about two-thirds of the core's radius. Below that, the core begins to freeze (i.e., the iron changes phase from a liquid into a solid) and remains solid to the center of the Earth. The temperature and pressure have been increasing this whole time as we've descended into the planet, reaching about 10,000 degrees Fahrenheit (that's hotter than the surface of the Sun) and 3.65 million bars of pressure. This increase in pressure is crucial in explaining a fun curiosity of the Earth's core: The temperatures are hottest at the center, but this is where the core starts to freeze. That's because the freezing point of iron depends more strongly on the pressure than on the temperature deep in the core.

We've now traversed the entirety of Earth (admittedly we have to get back out of the planet now that we're at the center) in our hypothetical vessel. Since no such vessel actually exists, scientists rely on indirect methods to learn about the inside of the planet. The coming chapters will discuss some of those methods as well as the fascinating results that they have revealed.

To study Earth's interior, it's crucial that we explore other planets as well. Aside from being really cool places, they offer insights to Earth's interior, as well as its formation and early history that we could not have discerned without them. So now let's get to know our nearest neighbors.

CHAPTER 2

Gazing Outward

ONE SUMMER MY PARTNER AND I drove the 700 miles from our home in Maryland to my parents' home in Sudbury for a vacation with my extended family. We frolicked all day in Ramsey Lake, swimming and feeding the ducks that were making their rounds. Then we prepared for one of our epic family dinners: Mom was in the kitchen cooking blackbirds and moose that she and my father had hunted. The blackbirds were served like tiny Cornish hens, stuffed with sage, butter, and bacon, and then slow-roasted to perfection, and the moose was cooked in a stew.

One of my nieces and I were setting the giant granite table in the sunroom when I noticed one of her fingers was wrapped in a bandage.

"What happened?" I asked.

"Oh," she said, "I got frostbite."

"In the summer?" I inquired.

"Well," she explained, "I had a pint of ice cream, but no spoon, so I used my finger."

We burst out laughing, and I said, "Only someone in this family can understand the pain we're willing to go through for good food."

We can learn a lot about ourselves from our family ecosystems. We naturally compare and contrast our talents, interests, and characteristics with those of our siblings and also look at our parents for answers to the questions of why we are the way we are. My sister, Maike (pronounced "mica," like the mineral), and I share a love of nature, learning, and creative arts, but she was a star athlete while we were growing up, and I could barely touch my toes or do a somersault. My sister Maria and I share a quick wit and rebellious streak, but her gregarious nature and ability to command a room just by walking into it was something I could only admire given my tendency to hide in corners and avoid conversation at parties. My blended family made it clear to me that some of these shared family traits were due to genetics, or "nature," whereas others were due to environment, or "nurture."

The solar system can be considered a family as well. The planets "grew up" together as they formed from the same solar nebula—a cloud of gas and dust particles that encircled the Sun as it was forming. There's an influence analogous to genetics here since all the planets, moons, asteroids, and other components of our solar system were made from the same building blocks and physical processes—the elements present in the solar nebula and the same set of physical laws that govern all matter. There's also a "nurture" element here, in that the planets experienced conditions such as temperature, solar wind, and impacts that affected their growth based on where they formed in the solar system.

This is excellent news if you want to understand the Earth, as it means that our planet has family members that you can investigate to learn more about it. In some ways, you can think of Earth and the other planets as sharing some DNA the way family members do, so looking at your sister's DNA test results also tells you something about your DNA. It also means we have other case studies to test our theories about how Earth works. If a scientist proposes a theory to explain a phenomenon on Earth, then they can test it by applying the theory to another planet. Here's my favorite example: We know Earth's magnetic field is important for shielding our atmosphere from solar wind stripping. That is, if it weren't for Earth's magnetic field, particles in the atmosphere would more easily blow off Earth because of collisions with energetic solar wind particles. The solar wind would sort of act like a giant hair dryer pointed at Earth. If we had only Earth to study, we might therefore conclude that a magnetic field is necessary to have a thick atmosphere on a planet, because without one, the atmosphere would eventually be completely blown off.

Luckily, we have data from other planets. Both the planet Venus and Saturn's moon Titan have thick atmospheres. Venus's atmospheric surface pressure is a crushing 92 bars—92 times Earth's surface pressure, which is similar to the pressure you would experience about half a mile deep in the ocean. Titan has a much more reasonable surface pressure of 1.5 bars. But neither of these bodies has a magnetic field. For Titan, this is because, as far as we know, it lacks a metal core; Venus has a metallic core, but either it doesn't convect, or it convects too slowly to maintain a dynamo. On the flip side, Mercury has a magnetic field, but no atmosphere. These counterexamples don't imply that magnetic fields aren't involved in maintaining an atmosphere, just that the process is more complicated, and other factors need to be included in order to understand the full picture.

Similar comparisons between planets help us to study other important questions such as, "Why does Earth have plate tectonics?," "What causes earthquakes?," and of course, the big one: "How did life form on Earth?" As a first step in these comparisons, we need to characterize each of the planets, and figure out what they're made of, how big they are, and what processes govern the features we can observe with various instruments, such as magnetic fields, volcanoes, and atmospheres. With that information, we then need some important context to frame our inquiry—like how did the solar system form, and what are the rules that govern planets? Essentially, we need to discuss the equivalent of "biology" and "evolution" for planets: What are planets made of, and what are the most important laws that govern the processes that create them?

HOW THE SOLAR SYSTEM FORMED

———

It all starts with the birth of our solar system. About 4.6 billion years ago, a portion of the molecular cloud in our area of the galaxy was compressed, possibly by shock waves emitted by a nearby star ending its life cycle through a supernova.[1] The compression caused gravitational collapse, in which the material in the region continued collapsing into itself, becoming denser and denser and forming a molecular cloud core. The center of that cloud core would eventually become our Sun.

The collapse happened quickly, taking "only" about 100,000 years for the region to shrink 100 times in size to its current extent of "merely" 20 billion miles across.

The composition of the molecular cloud core ultimately determined the composition of everything in our solar system. As the name suggests, it was mainly built from molecules. The most abundant molecule in the universe, molecular hydrogen (H_2), existed as a gas in the molecular cloud core. The cloud also contained smaller amounts of other gases and some tiny solid particles that astronomers call "dust." The dust was actually made of the remnants of dead stars. All elements heavier than lithium are made inside stars during nuclear fusion. After a star ends its life cycle, it might go through a phase (for stars a few times bigger than the Sun, sometimes a supernova) during which it sheds its outer layers. That material litters nearby regions and seeds the growth of dust grains. The dust grains

Solar System Formation

4.6 billion years ago, our solar system was part of a wispy cloud of gas and dust

Part of the cloud collapsed in and formed a flat spinning disk of dust and gas

The center collected enough matter for nuclear fusion to begin—our Sun was born

The material in the disk began clumping together into larger pieces to form planetesimals, planetary embryos, and eventually planets

Present day:
Our solar system now is comprised of the Sun, eight planets, many moons, dwarf planets, asteroids, and comets

present in our molecular cloud core eventually formed all the solids in the solar system, including you and me. Carl Sagan perhaps said it best: "The nitrogen in our DNA, the calcium in our teeth, the iron in our blood, the carbon in our apple pies were made in the interiors of collapsing stars. We are made of starstuff."

There is another very important component in the gestation, without which planets would never have formed: rotation. The material surrounding the center of the collapsing cloud core had a wee bit of average rotation around the center. All the material was moving around, and most of it was in random directions, but if you averaged all the moving directions, they didn't all cancel out. Gravity acted to move material to the center of the cloud core, creating the "proto-Sun," but this rotation created a centrifugal force that could partly counteract the gravity. This caused some of the collapsing material to form a disk around the collapsing core, which we call a protoplanetary disk (for planetary taxonomy, see pages 43–46)

This slightly spinning protoplanetary disk continued to contract as the proto-Sun grew and its rotation sped up due to a law of physics known as "angular momentum conservation." Angular momentum is a measure of how much spin you have and where that spin is (i.e., how extended into space the object is that's spinning). If there are no external forces acting on an object, it must conserve its angular momentum—the product of its spin rate and its shape can't change.

I was intuitively familiar with this conservation law long before I learned it in physics class. The weather in Sudbury shaped my childhood, ensuring that winter ice- and snow-related sports were an intrinsic part of the culture. Almost everyone I knew either played hockey, ringette (a sport invented in northern Ontario that's sort of like hockey but involves a straight stick and a ring that you maneuver over the ice with it), or was a figure skater or skier. My parents were curlers, which was the only sport I played competitively in high school (admittedly, I was the team alternate). The important thing about ice-related sports is that the ice has such low friction that if you try to spin on the ice (like in figure skating), you're subject to angular momentum conservation.

I got to test out the low-frictional behavior of ice again recently. In the winter of 2022, Johns Hopkins University installed an outdoor ice rink for free use by the community. As soon as I heard about it, I hurried to contact a group of colleagues in my department who had previously demonstrated a willingness for quirky excursions, like floating down the Gunpowder Falls in inner tubes. Everyone agreed to come. We reserved a timeslot on the ice, left the Earth and Planetary Sciences Department building with all the constantly growing piles of work we each had gnawing at us, and headed over to the rink. I put on orange rental skates and eagerly approached the ice. That's when panicked thoughts began to run through my head: Is skating like riding a bicycle? Will it matter that I haven't put on a pair of skates in almost 30 years? How horrible will falling on the ice in my 46-year-old out-of-shape body feel?

I tested one blade on the ice while clutching the nearby outer fencing, then I put down the other skate. Balance was suddenly an entirely new concept and I felt like a baby trying to stand for the first time. "Maybe it's just because I'm not moving?" I foolishly thought. I attempted a bit of motion and quickly lunged for the fence again, resigned to the fact that skating is most definitely not like riding a bicycle. I could feel myself losing serious Canadian credibility points the whole time and I expected some official (maybe in a Mountie uniform) to come by and take my Canadian passport away from me. But ultimately I was reminded that the low friction of ice allows for some interesting laws of physics and motion.

I've now temporarily resigned myself to watching ice-related sports on TV rather than playing them. My favorite winter sport is figure skating. I grew up during the "Battle of the Brians," when Brian Orser (Canadian) and Brian Boitano (American) battled for top spots in each Olympic or world championship. Similarly, but on an entirely different scale than the protoplanetary disk that surrounded the Sun, rotation is important for figure skaters. In both the jumps and the spins, it's all about completing the turns, and if you want more turns, you have to spin faster. Figure skaters know how to do this. Skaters may begin their rotations with their arms out, but then they draw them in close to their body. This causes them to spin faster because of angular momentum conservation. It's a fun concept to test out. You can do it yourself on the ice (if you dare) or in a good rotating office chair, which is somewhat safer.

Angular momentum conservation was also an important factor for the protoplanetary disk that encircled our forming Sun. The protoplanetary disk was a massive structure of dust and gas that formed the building blocks of all objects in our planetary system, from meteors to gas giants. The small amount of average rotation possessed by the disk meant it had angular momentum. But gravity caused the disk to continuously contract. As particles in the disk moved inward, angular momentum conservation meant their rotation had to speed up. This is why only a tiny bit of rotation was needed in the beginning. The same is true for the formation of other stars. We can surmise that any forming star with a contracting nebula around it will become a protoplanetary disk with gas and dust rotating around the star rather than all the material falling into the star.

The physics of this makes total sense. Although we can't return to the dawn of our own solar system and check whether this is what actually happened 4.6 billion years ago, we can look outward to where other stars are forming. I like to think of this as looking at other families to see how unique or weird your own is. Here we learn that the solar system family started like many other stellar families. With telescopes like Hubble, JWST, and Spitzer (the former two administered by the Space Telescope Science Institute, which is affiliated with Johns Hopkins University), we see the birth of other star systems and we can witness collapsing molecular cloud cores throughout the galaxy in the form of rotating disks of gas and dust surrounding their central stars.

But how did we go from a disk of gas and dust to a few isolated planets? As material continued to fall into the center of the cloud core, creating the proto-Sun, it produced a lot of heat. The proto-Sun thereby heated the protoplanetary disk, resulting in hotter temperatures closer to the Sun and cooler temperatures further out.

Just like ice cream is solid if left in the freezer and a disappointing liquid if accidentally stored in the fridge (am I the only one who's done this?), the compounds in the solar nebula were in whatever phase was appropriate for the temperature and pressure of their environment.

Closer to the center of the protoplanetary disk, where the temperatures were hotter, only materials with high melting (or equivalently, freezing) points were able to condense—namely, rocky compounds (made up mostly of silicate and magnesium minerals) and metals. But further from the proto-Sun, as the temperature continuously decreased, volatile compounds made of combinations of hydrogen, oxygen, carbon, and nitrogen, like water (H_2O), carbon dioxide (CO_2), and ammonia (NH_3), began condensing along with the rocks and metals. These condensed materials became the building blocks of planets.

Planetary formation required that these condensed dust grains had to find their way from micron-sized specks to become planets that are thousands of miles in radius. When the dust grains were tiny, their masses were too small for gravity to have the attracting power needed for them to grow effectively. Instead, another force was involved in getting the tiny dust grains to cluster together and then grow.

If you've ever rubbed a balloon against your clothing and then stuck it to your hair or a wall, or suffered the annoying shock and spark that results when touching a door handle after dragging your feet on a carpet, you've experienced this force—the electrostatic force. This force acts between objects that have an electric charge, either positive or negative. Like charges repel, and opposite charges attract. The balloon is stuck to your hair because rubbing it on your clothes adds electrons, which have negative charges, to the balloon. The wall or your hair are then more positively charged than the balloon and so they attract. Similarly, when you rub your feet on the carpet you're adding negatively charged electrons to yourself. As your finger approaches the door handle, the excess electrons are attracted to the more positively charged door handle, and they create a current to travel to the door handle—the spark—and thereby rebalance the electrical charge between the two.

The dust grains in the protoplanetary disk had charges as well. The attraction of oppositely charged dust grains allowed them to stick together. As more and more stuck, they grew into larger and larger bodies. Once they reached about an inch wide, gravity could act effectively to stick them together, giving them cohesion. We refer to them as "pebbles" rather than "dust" at this stage. Once these pebbles were hundreds of yards to miles wide, gravity became the main force holding them together. At this stage, we call the objects "planetesimals."

Planetesimals orbited the proto-Sun and grew by gravitationally attracting the nearby dust and pebbles, but they were also affected by collisions with other planetesimals. The solar system was a busy place at this time, and close encounters by two planetesimals could cause a variety of things to happen. If the relative speed between two colliding planetesimals was slow, they coalesced to form a larger planetesimal. If the speeds were higher, then their collision could break the bodies apart. A near-miss could also occur where the planetesimals didn't actually collide, but the gravitational acceleration they experienced from the encounter flung them into wacky orbits in the solar system, and perhaps on collision courses with other planetesimals. Near misses could have also flung planetesimals out of the solar system, or very far away from their original locations. Indeed, like in many families, there are members of the early solar system now hurtling off at great distances, never to return home. We've also detected visitors from other stellar systems that happen to pass by ours while on a similar journey, like the interstellar object 'Oumuamua observed in 2017.[2]

Ultimately, some planetesimals continued to grow and cleared out their neighborhoods by either accreting the other planetesimals in the area or flinging them far away. Planetesimals that grew to about moon-sized (~1,000 miles wide) are called planetary embryos. Eventually, planetary accretion left us with the handful of planets we have today. Let's examine these planets more closely.

TERRESTRIAL PLANETS

We'll start by looking at our closest solar system family members (both figuratively and literally): the "terrestrial," or "rocky," planets: Mercury, Venus, and Mars. Let's also include the Moon here since it's another close-by body with similar characteristics. Indeed, the only reason a body is classified as a "moon" has to do with the fact that it orbits around a planet rather than orbiting the Sun. So, the composition, structure, and dynamics of moons are very similar to those of planets, and there's little reason to consider them separately. Mercury, Venus, Earth, Mars, and the Moon are all quite comparable in composition and structure. They have rock-rich crusts and mantles, and iron-rich cores. They come in a variety of sizes, with Venus and Earth quite similar, Mars about half the radius of Earth, Mercury about a third, and the Moon about a quarter. There are differences among their interior compositions and structure, and these can be attributed to both environmental

The Solar System Family Tree

Our solar system is home to more celestial bodies than the Sun and its planets—it also contains moons, asteroids, comets, and more. Here's a user-friendly guide ("translated" from more complex scientific definitions), from the smallest particles to the largest bodies, that reflects the variety of Earth and its "relatives."

Dust: Small, solid particles—silicates, carbonates, ices, and more—that are the building blocks of other materials and space objects. They can coalesce into rings around planets such as Saturn and Uranus through gravity, or exist in clouds several light years across, as happens in nebulae. The smallest dust particles could be just a few molecules put together, while others are more like a grain of sand. Space missions like NASA's Stardust (1999–2006) have gathered samples to return to Earth. Stars also light up dust and can be seen on Earth as zodiacal lights, as in images from the JWST and Hubble Telescope.

Rock: An object that has aggregated out of multiple rock-forming materials. In space, rocks may be gathered out of dust pulled together by various forces (including gravity) or could be broken-off chunks of other space objects. By astronomical definitions, some could even be liquid. They could also be made out of ices, depending on where they formed. On Earth, geologists define rocks as naturally occurring combinations of solid minerals that are fused together. On planets, they can form in several ways, including from volcanic eruption or through sedimentation of particles in liquids.

Comet: An object made of rock and ice that at times has a tail, especially the nearer it comes to the Sun. These are also leftovers from the formation of the solar system, albeit from a point far enough from the Sun where solid ice forms more readily (hence, also, the number of icy moons in the outer solar system). Some take just a few years to orbit the Sun, while others can take tens of thousands of years. One of the most watched for is Halley's Comet, which returns to Earth's vicinity approximately every 76 years. (Its next visit is in 2061.)

Asteroid: Space rock that's left over from the formation of the solar system. Many are small and irregularly shaped; Ceres is the only known asteroid large enough to be rounded by its own gravity, thus also making it a dwarf planet (see definition below). NASA has sent a handful of missions to asteroids, including the OSIRIS-REx mission, which scooped a piece of asteroid to return to Earth, and the Double Asteroid Redirection Test (DART) in 2022, which smashed into a small moon of an asteroid to test possible defense strategies if one came near Earth.

Meteor: An asteroid (usually, but not always, small) or small comet that enters Earth's atmosphere. It's termed a meteoroid when it breaks through the atmosphere and a meteorite when it survives entry and is found on Earth's surface.

Planetesimal: Space objects ranging between 1 and 100 miles in length that become the building blocks of protoplanets. Many asteroids and comets may be planetesimals that never formed into larger bodies.

Protoplanet: A space object built out of planetesimals. Most are the size of a medium to large moon (about 100–1000 miles across). Some astronomers consider Ceres to be a leftover protoplanet, along with some of the larger asteroids. Protoplanets form the building blocks of planets.

Moon (or satellite): An object in orbit around another planet. Some satellites may be the size of planets, whereas others may be as small as an asteroid or comet. There is currently no defined limit to how small a moon can be. Earth's moon is believed to have formed out of the debris of a collision between early Earth and a protoplanet called Theia, and could be a mixture of materials from both worlds. While we have only one moon, Jupiter has at least 95 (including Europa, which is believed to have a vast sea underneath its frozen surface) and Saturn has at least 145 (including Enceladus, which also appears to have an ocean under its ice-covered surface).

Dwarf planet: As defined in a 2006 International Astronomical Union (IAU) decision, an object in orbit around the Sun that is nearly round due to internal forces, but does not clear out other objects in its path. The term was coined when a number of objects the same size as Pluto were found in the same region, clustered a bit like the asteroid belt, and called the Kuiper Belt. This is when Pluto was reclassified. Many of those objects are made of ice surrounding rock. Interiors are often planet-like, although composition can vary wildly. Ceres is also a dwarf planet, though it was once designated as a planet.

Planet: This is tricky. The IAU defines a planet as an object orbiting the Sun that's nearly round due to internal forces and is the only large object following that specific orbital path ("clearing the neighborhood," which so-called dwarf planets do not). But this definition isn't one-size-fits-all; planetary surfaces and interiors encompass a diversity of forms and materials. But by the IAU definition, planets outside our solar system aren't planets; a planet-like body outside our solar system is called an exoplanet. Planets can be rocky (terrestrial), made of ices of heavy elements (ice giants), or be predominately made up of hydrogen and helium gas (gas giants.) Exoplanets a few times larger than Earth are called super-Earths. Gas giants that closely orbit their stars are called Hot Jupiters.

Star: A ball of hot plasma that fuses hydrogen into helium, which usually starts at temperatures above 4,000,000 Kelvin. The Sun is a star, one of an estimated 100–400 billion in the galaxy. Some are small and dim, while others would engulf most of our solar system. For more on what happens to stars after they die, see chapter 7.

FURTHER READING

Broad, William J. "Flecks of Extraterrestrial Dust, All Over the Roof." *New York Times*, March 10, 2017.

Double Asteroid Redirection Test: "NASA's First Planetary Defense Test Mission." Accessed February 28, 2023. https://dart.jhuapl.edu/.

Howell, Elizabeth. "Europa: Facts About Jupiter's Icy Moon and Its Ocean," Space.com, October 26, 2022. https://www.space.com/15498-europa-sdcmp.html.

International Astronomical Union. "Pluto and the Developing Landscape of Our Solar System." Accessed February 28, 2023. https://www.iau.org/public/themes/pluto/.

effects and the random unique events each has encountered throughout its particular history.

An intriguing comparison is in how much "core" each of these planets has. Iron cores were a by-product of the violent processes that occurred during planet formation. All those impacts of planetesimals added heat to the forming planets and increased their temperatures. Growing planets also had radioactive isotopes in their interiors. As those decayed, more heat was generated in the planet's interior. If a planet's accretional and radiogenic heating raised the temperature beyond the melting point of the rocks and iron, then the planet rearranged itself. The densest materials, mainly iron, fell to the center of the planet, creating the core. The lighter rocky materials floated to the top of the planet (buoyancy at work again). We call this separation into different layers of rocks and iron in a planet "differentiation," and it's why the terrestrial planets all have "cores" and "mantles."

For Earth, the iron-rich core makes up about 55% of the radius of the planet, which translates to about 15% of the planet's volume. Mars and Venus are similar. We think this similarity is not coincidental, but that it's instead the result of these planets having formed in the same part of the protoplanetary disk and hence from similar planetesimals. That means that Earth, Mars, and Venus received similar fractions of the elements that make up the mantle and the core during their formation.

But like every family, the rocky planets have two outliers: Mercury and the Moon. Whereas the Moon's core is only 20% of the radius (less than 1% of the volume), Mercury's core makes up a whopping 85% of the radius (about 60% of the volume). We have theories as to why, and interestingly, the main player in explaining both Mercury and the Moon is the same: a giant impact.

For Mercury, the leading theory involves the planet having started life looking very different from its current self. Proto-Mercury could have been about twice as large—with a typical-sized mantle on top of that core.[3] It may even have been bigger than Mars today. But then something catastrophic happened. Evidence suggests that a violent impact occurred between proto-Mercury and another protoplanet, stripping off most of Mercury's mantle, leaving the iron-rich remnant we see today.

Although the culprit for explaining the Moon is the same—a giant impact—the details are different. The leading theory, first set forth by William K. Hartmann and Donald R. Davis in 1975[4] and refined over the subsequent decades, involves a Mars-sized body colliding with Earth. That object, called Theia by later planetary scientists, had grown large enough and hot enough that core differentiation had occurred on Theia, much like it had on Earth. That collision vaporized Earth's surface and flung a bunch of the crust and mantles of both Earth and Theia off the planet. Some of that material remained in orbit around

Earth and, over about 40 years, that material coalesced into larger and larger bodies that eventually formed the Moon. I'm constantly shocked by how short the planetary formation process time periods are, given that most events in our solar system's history are referenced in millions and billions of years. Basically, if humans had existed back then—and somehow survived this giant impact—a parent could have told their kids about the "era before Earth had a moon" and how different it was back then (while their teenagers roll their eyes at having to hear this lecture once again). Since the original building blocks for the Moon were mostly material from Earth's and Theia's mantles, there would be much less iron available to form the Moon's core.

Giant impacts like this aren't far-fetched. We know that impacts happen all the time in our solar system. Every meteor that shoots across the sky and reaches the surface is an impact. Nowadays, the bodies impacting us are, luckily, not that big since there isn't much large material left in the solar system. But early on, fierce impacts with planetesimals and planetary embryos happened more often. There are scars of some of these on the terrestrial planets in the form of impact craters. All the giant circular features on the Moon, visible on a clear night, are impact craters. But impact craters were produced by small enough impactors that they didn't entirely destroy the planet's surface. Larger impactors wouldn't have left any crust to preserve the record of the event. It's these larger impactors that led to major planetary changes, like the formation of Earth's Moon and the removal of most of Mercury's mantle.

GIANT PLANETS

The terrestrial planets—versus the gas and ice planets—share the most similarities to Earth, but they make up a miniscule fraction of the mass of all the planetary bodies in the solar system. The solar system is really dominated by the giant planets, found farther out in the solar system: Jupiter, Saturn, Uranus, and Neptune. I consider these planets akin to more distant relatives of Earth, although still very much part of the solar system family. It might be hard to see any similarities between, say, Jupiter—a giant ball of hydrogen with a mass of over 300 Earths—and Earth, but there are some. Beyond these four giants, are an assortment of hundreds of small objects like Pluto that don't rise to the current definition of a planet as put forth by the International Astronomical Union. (Pluto was demoted in 2006.) While they're in orbit around the Sun they're either not round because of internal forces, or they don't "clear the neighborhood" of other objects, meaning that a number of them share similar orbits.

From the genetics/nature standpoint, the building blocks in the protoplanetary disk were different farther from the proto-Sun. In addition to the condensed rocks and metals, the lower temperatures in the outer solar system allowed volatile compounds like water, ammonia, and methane to condense and become part of the building blocks of planets. Planetary scientists call these volatile compounds "ices." Note that "ice" here isn't meant to indicate just frozen water, but instead, a compositional group: molecules made from combinations of hydrogen, nitrogen, oxygen, and carbon. More building blocks in the outer solar system meant bigger planets, and this partly explains the size difference between the terrestrial planets and giant planets.

The other big difference is a consequence of the giant planets' larger sizes. The solar nebula didn't just have rocks, metals, and ices. It also had gases—most importantly, a whole lot of hydrogen and helium. But gases were harder to capture gravitationally than solids. Planets needed to have a critical mass (about ten times Earth's mass), in order to gravitationally attract the gases in the early solar system. The giant planets grew big enough, fast enough, to do this. However, they didn't all end up the same, and there are some major differences between the giant planets.

All four giant planets are composed of similar amounts of rocks and ices, around 10–15 Earth masses worth. The rate at which planets accumulated the rock and ice building blocks was related to how quickly they encountered material to accumulate. Farther from the Sun, the protoplanetary disk was less dense (so the material was spread further apart). Planetary orbits were also slower further from the Sun. Jupiter and Saturn, which were closer to the Sun, therefore likely grew faster than Uranus and Neptune. It appears their growth was fast enough for Jupiter and Saturn to capture significant amounts of gas in the solar nebula and end up as "gas giants." In contrast, Uranus and Neptune may have taken longer to grow to critical mass. The gas in the protoplanetary disk stuck around for only about 10 million years, so by the time Uranus and Neptune grew big enough to capture any, there wasn't much left. They are therefore referred to as the "ice giants" because they are mostly made up of icy building blocks. I always picture Uranus and Neptune arriving late to the dinner table and there not being any food left for them because Jupiter and Saturn gobbled it up—poor deprived planets! This is the leading theory as to why Uranus and Neptune have thinner gas layers atop their ice, rock, and metal interiors than Jupiter and Saturn, but there are still issues with this theory.[5]

We also think Uranus and Neptune both formed closer to the Sun than their current locations. In a theory known as the "Nice" model (named for the city in France where it was developed, not a judgment on the theory's pleasantness), all four giant planets formed in a more compact configuration and then migrated around a bit through gravitational interactions, eventually ending up in their current orbital distances.[6] There's also some evidence that the solar system might have had a fifth giant planet early in its history. When scientists create simulations of the formation of the solar system, it's really hard to get the giant planets to eventually end up in their current orbits unless there is a fifth giant planet that eventually gets kicked out of the solar system from these gravitational interactions.[7] The peculiar orbits of some comets out in the Kuiper Belt beyond Neptune also hint that there might even be an additional planet that we haven't yet discovered out there today! It's typically called "Planet 9" (sorry, Pluto) and evidence suggests it's about 5–10 times as massive as Earth (so a bit smaller than Neptune).[8]

Although the giant planets have characteristics quite different from those of the terrestrial planets, I like to think of them as being terrestrial planets in their hearts and just having a lot more ices and gases piled up on top of them. The reality is not that simple, and we'll discuss some complications in later chapters.

SOLAR SYSTEM LEFTOVERS

During the spring and summer of 2020, with the COVID-19 shutdown and work-from-home requirements of many employers, I found myself trying to find new things to do at home to battle the terror and sadness of what was unfolding around me. Like many others who were striving for a small escape, and perhaps some ability to nurture and grow something, I turned to the frustrating art of maintaining a sourdough starter for bread making. In the initial stages of creating a sourdough starter, you need to feed it twice a day and pray that it eventually starts the growth and expansion phase to become a "mature" starter. As someone who never had children, I probably became way too attached to my starter, whose name was Hercule (after the great literary detective Hercule Poirot). I was devastated when my partner had to sit me down one day to break the news that three-month old Hercule, whom I had left in the oven one morning after feeding, had accidentally been baked when my partner preheated the oven for a frittata.

Happily, there were future sourdough starters, and bread making was attempted. I love the chemistry involved in this process. You take materials that are in a particular form—the flour, water, salt, yeast—and by combining them, you allow chemistry to take over to produce a completely different end product: delicious, crusty bread with a chewy center. Planets are kind of like bread in this way. The final product is the result of a bunch of ingredients coming together that then experience a variety of chemical and physical reactions resulting in a final product not necessarily recognizable as the original ingredients.

If someone didn't know what bread was made from, they could discern information from the clues left in my kitchen. These are the "leftovers" from the bread's formation like the splatters of flour or spilt salt on the floor, or the sticky yeast in the jar near the sink that I dread cleaning. The solar system has leftovers from its formation as well: asteroids, comets, and dwarf planets were left over from the early solar system, and some moons formed from leftover material around freshly minted planets. The moons in the outer solar system provide some intriguing case studies for planet formation. For example, the Galilean moons that orbit Jupiter—Io, Europa, Ganymede, and Callisto—formed from a disk around Jupiter in a manner similar to how the planets formed from the protoplanetary disk surrounding the Sun. Saturn's largest moon, Titan, might also have formed this way. However, there are hundreds of other moons in the solar system orbiting the planets. They are mostly smaller than Titan and the Galilean satellites, and many of these were planetesimals that never ended up getting accreted inside a planet and were eventually captured into orbit around a larger planet. That's ultimately what most comets, asteroids, and dwarf planets are as well—the leftover building blocks that didn't become parts of planets.

Many of these moons, comets, and asteroids are fascinating to study as unique bodies themselves, but they also provide us with evidence of the materials from which Earth and the other planets were made. One of the major pieces of evidence that the dense material at the center of Earth is made mostly of iron comes from the study of meteorites. We can see that a specific class of them, the "iron meteorites," are samples from planetesimals that had been shattered by an impact and that parts of their cores had fallen to Earth as meteors. In a later chapter we'll further discuss how we learn about planetary interiors from meteorites.

EXOPLANETS: MEETING OTHER FAMILIES

———————

Of course, our family of planets aren't the only ones out there—
we know of thousands in the Milky Way alone. The confirmed
discovery of the first extrasolar planets (also called exoplanets, with "exo-" meaning "outside," for outside the solar system) in the 1990s triggered a rethink on how planets form.

Astronomers had been looking for exoplanets since the
nineteenth century. The first-ish claim of an exoplanet by an
astronomer came in 1855, when William Stephen Jacob of the
Madras Observatory in present-day India claimed to see anomalies in the orbit of the binary star system, 70 Ophiuchi, that
indicated a third object in there.[9] Three other claims would be
made in subsequent decades about 70 Ophiuchi without turning up much. In the mid-1960s, a Swarthmore astronomer
named Peter van de Kamp claimed that a nearby star named
Barnard's Star had at least one planet around it, but it ended
up being an error with the observatory telescope.[10]

All of these observations came from trying to see how the
positions of stars changed visually from a planet pulling on
them. Finding a planet with this method is *really* tough. In the
1970s, Canadian astronomers Bruce Campbell and Gordon
Walker of the University of British Columbia finally came up
with a good way to find planets: break down a star by its light,
and watch how the spectra of that star changes as the velocity
of that star changes relative to us due to an orbiting planet.[11]

The amount of change is small but detectable with the right instruments; Campbell and Walker made the breakthrough discovery of the right gas to put in the instrument to watch the way the spectra shifted toward or away from Earth in tiny amounts. This method is called the radial velocity method of exoplanet detection. It's a lot easier than trying to watch minor deviations in the position of a star, and by the late 1980s, there were a few strong candidates, but they weren't confirmed. In 1988 Campbell, Walker, and Stephenson Yang of the University of Victoria identified an exoplanet called Gamma Cephei Ab,[12] but later retracted their discovery only to have it reaffirmed several years later.[13]

The first crack in the case came in 1992, when astronomers tuning into the dead, dense core of a star that went supernova noticed periodic blips in its radio signal, which are normally like an ultrafast metronome. The astronomers soon realized that it was due to planet-sized objects orbiting them. But these planets are perhaps the weirdest known, as the star system had already been obliterated in the supernova, and these "planets" may have been formed from the corpses of the first planetary system. They didn't tell astronomers much about how planets might form around a Sun-like star.

Until the early 1990s, scientists used a combination of observations, experiments, and computer simulations to develop an understanding of how our solar system formed and evolved to its current state. Many of the important processes that governed the formation of our solar system were considered to be broadly applicable and hence were used to infer properties of planets that we might eventually find orbiting other stars. For example, the fact that the temperature structure of the protoplanetary disk resulted in rocky planets forming closer to the Sun and giant gas or ice planets forming further away seemed to be a principle that could be applied to other stellar systems as they formed planets.

It was, therefore, somewhat of a shock when, in 1995, scientists discovered that one of the first confirmed exoplanets around a sun-like star was a giant planet about half the mass of Jupiter, orbiting about eight times closer to its star than Mercury does our Sun. This planet, discovered by Didier Queloz and Michel Mayor of the Geneva Observatory in Switzerland and called 51 Pegasi b,[14] orbited its home star in just four days. (Queloz and Mayor received the Nobel Prize in Physics for this discovery in 2019.). The duo were one of a handful of teams combing through radial velocity signatures of nearby stars looking for the presence of a planet. According to our theories of planet formation, a giant planet forming that close to its star should not have been possible. In the decade that followed, more giant planets were discovered orbiting very close to their stars, and it became clear that a major process was missing in our understanding of planet formation, mainly because it had not happened in our own solar system.

We now understand that the missing process is "planetary migration," and it's a result of gravitational forces. As discussed earlier in this chapter, gravity was a major player in the formation of our solar system through the resulting collisions and close encounters experienced by growing planetesimals. But gravitational interactions can also be quite nuanced. For example, as a planet forms, it can interact gravitationally with the gaseous disk in its vicinity, creating waves in the disk that then change the planet's orbit. This can cause planets to drift inward or outward in the protoplanetary disk, depending on the details of the situation. Planets can also experience resonant gravitational forcing with other planets when their orbits have periods that are multiples of each other. These migration processes can result in very different final locations for planets compared to their starting locations (like what happened with Uranus and Neptune), and it's the main cause of why we find many giant planets very close to their parent stars. Once 51 Pegasi b was found, astronomers began to find more gas giants, many of them in close-in orbits.

It wasn't until the mid-2000s that smaller, rocky exoplanets were discovered, and it wasn't until the 2010s that planets were discovered, sometimes in batches of hundreds or thousands, thanks to a technique called the transit method. This ingenious method waits for a planet to pass between its star and Earth. It takes a planet aligned just right, but if you send up a telescope like Kepler or TESS that can stare at a bunch of stars at once, you can witness a lot of these tiny blots. A planet might change the output of its star's light just a little, but it's enough for astronomers to tell roughly how big it is. It's effective for catching a lot of planets in a particular amount of time, and for telling us how big around they are, but not for telling us how massive they are or what they're made out of.

As of 2023, more than 5,300 exoplanets have been discovered. They range in size from about one-third the radius of Earth to a few times the radius of Jupiter. Some of these planets are quite different from those found in our solar system. There are "super-Earths": rocky planets with masses up to ten times that of Earth; "mini-Neptunes": smaller versions of our Neptune, and "Hot Jupiters": Jupiter-mass planets with much larger radii because they've been puffed out due to their extreme temperatures.

Some exoplanets are found in systems or families (i.e., multiple planets orbiting their central star). One of the most exciting families is the TRAPPIST-1 system, located 40 light-years away in the Aquarius constellation. Its star has seven observed orbiting planets. The planets are all terrestrial-sized, with the smallest being about Mars-sized, and the largest a bit bigger than Earth. Recent research suggests the innermost planet has no atmosphere, but planetary scientists are holding out hope that the others could have atmospheres and water—and a few could even be habitable if that's the case. That means the temperatures on these worlds might be within range for liquid water to be viable on their surfaces, a crucial component for the formation of life as we know it.

Based on size projections of some planets and estimates of data like their home star's temperature and size, we can say that a planet is in what's called the habitable zone, also sometimes called the Goldilocks zone. An astronomer on an alien planet who managed to find the eight known planets in our solar system could safely say that Mercury is too hot to host life as we know it, and Jupiter on out are too cold for life, but there's a hitch: they'd find three planets in our Sun's habitable zone: Venus, Earth, and Mars. That's where the interiors and atmospheres of the planet matter as much as the activity of their host star and the materials present at their formation.

As a demonstration of life imitating art, some exoplanet equivalents of famous sci-fi planets have also been discovered. Kepler-16b, an exoplanet that orbits two stars at once, has been likened to Tatooine (Luke Skywalker's home world in the Star Wars franchise) because two sunsets would be visible from the surface, while a particularly cold super-Earth called OGLE-2005-BLG-390Lb was given the name Hoth for an ice planet in *The Empire Strikes Back*.

Arguably, the most intriguing question about exoplanets is: will we find another Earth? That is, a planet about our size with liquid water and an oxygen-rich atmosphere teeming with life. The search continues—with direct relevance to this book. To determine if we can find another Earth, first we have to look at what defines Earth. To examine that, we need to understand the chemical and geological processes that make our world what it is. Let's dive in a little into how planetary scientists investigate these processes both on Earth and on nearby planets.

Planetary Gestation

Billions of miles from Earth, spinning in the dimly lit, −370 degree Fahrenheit Kuiper Belt—a vast realm of "planetary leftovers" fossils from the solar system's formation—beyond Neptune's orbit, is an orange, flattened, 22-mile long snowman. Its name is Arrokoth, and it's the most distant object ever visited by a spacecraft—at least one of earthly origins.

This grumpy-looking body isn't really made from snow, but it is covered in methanol ice; in this frigid region, substances that would be in a gas form on Earth exist only in frozen states. Arrokoth is a 4.5 billion-year-old example of one type of planetary evolution, stunted in its early stage of gestation, presumably because it exhausted nearby raw materials to keep growing. Known as a "primordial contact binary" and a "planetesimal remnant," its two aggregated lobes formed from swirling nebular pebble clouds that merged slowly and gently through a spiraling gravitational attraction that slowly disconnected from the surrounding gases to form solid masses. Its "neck" shows no signs of stress fractures from a more volatile collision.

First detected in 2014 by the Hubble Space Telescope, whose operations center is on the Johns Hopkins campus, NASA's New Horizons mission photographed Arrokoth on New Year's Day 2019. This look at such a distant celestial body is a shining technological and human achievement. Hal Weaver, the former New Horizons project scientist (and current Hopkins professor and principal professional staff member at the Applied Physics Lab), says it best: "It's kind of mind-blowing that the spacecraft could successfully be targeted to

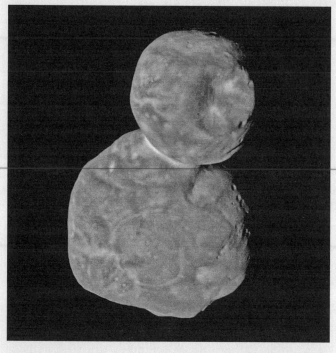

Arrokoth, discovered on June 26, 2014, photographed January 1, 2019. Image courtesy NASA / Johns Hopkins APL / Southwest Research Institute / Roman Tkachenko.

a point so close to this tiny body—just 2,194 miles away—after traveling 4.1 billion miles to capture these images while rushing past at a speed of 32,000 mph!"

While NASA often names moons, planets, and missions after Roman mythological figures (including the upcoming Artemis I and II missions), the New Horizons team proposed the name *Arrokoth*, the

Algonquin word for "sky," to the International Astronomical Union and Minor Planets Center, the authority for naming Kuiper Belt objects. The Virginia Panumkey tribe consented, and celebrated the choice during a naming ceremony.

FURTHER READING

Grady, Monica. "Why the Most Distant Object Ever Visited Looks Like a Snowman: Flyby Delivers Results," *The Conversation,* February 14, 2020. https://theconversation.com /why-the-most-distant-object-ever-visited-looks-like-a-snowman-flyby-delivers -results-131797.

Johns Hopkins Applied Physics Lab. "New Horizons: NASA's Mission to Pluto and the Kuiper Belt—About Arrokoth (2014 MU69). Accessed April 19, 2023. http://pluto.jhuapl .edu/Arrokoth/Arrokoth.php.

McKinnon, W. B., et al. "The Solar Nebula Origin of (486958) Arrokoth, a Primordial Contact Binary in the Kuiper Belt." *Science* 367, no. 6481 (February 13, 2020). https://doi.org/10.1126/science.aay6620.

NASA. "New Horizons: Far, Far Away in the Sky: New Horizons Kuiper Belt Flyby Object Officially Named 'Arrokoth,'" November 12, 2019. https://www.nasa.gov/feature/far-far -away-in-the-sky-new-horizons-kuiper-belt-flyby-object-officially-named-arrokoth.

CHAPTER 3

Telltale Planetary Parcels

NO DOUBT ABOUT IT—holding a meteorite in your hand—an alien solar system morsel—is a wild experience. It might have come from the Moon, Mars, or asteroids in near-Earth orbits.* Aside from their coolness factor, meteorites are arguably the most important tool for learning about what Earth and the other planets are made of and how they formed. While I was in college, one Martian meteorite caused a huge stir.

In 1996, the media and research worlds were abuzz. A 6.6-inch, greenish-gray rock from Mars, discovered in Antarctica in 1984 by Antarctic Search for Meteorites researchers, contained structures inside it that, to some scientists, looked like the sorts of things fossilized bacteria might leave behind. The rock, called Allan Hills 84001, was also one of the oldest fragments of another planet we have in our possession—a pristine

---

*Admittedly, the earnest side of me wants to point out that every single rock on Earth, not to mention every atom in your body, is made from material from space, too, so it's all cool, but buzzkill Sabine never wins this argument when I'm actually holding a meteorite.

relic from early Mars. President Bill Clinton, in an August 1996 address, said:

> Today, rock 84001 speaks to us across all those billions of years and millions of miles. It speaks of the possibility of life. If this discovery is confirmed, it will surely be one of the most stunning insights into our universe that science has ever uncovered. Its implications are as far-reaching and awe-inspiring as can be imagined. Even as it promises answers to some of our oldest questions, it poses still others even more fundamental.[1]

So . . . how did Allan Hills 84001 get here? Clinton's speech summarized it nicely:

> It is well worth contemplating how we reached this moment of discovery. More than 4 billion years ago, this piece of rock was formed as a part of the original crust of Mars. After billions of years, it broke from the surface and began a 16 million-year journey through space that would end here on Earth. It arrived in a meteor shower 13,000 years ago. And in 1984, an American scientist on an annual U.S. government mission to search for meteors on Antarctica picked it up and took it to be studied. Appropriately, it was the first rock to be picked up that year—rock number 84001.

Allan Hills, for which the meteorite is named, is a ridge of hills in the Victoria Land region of the continent. The area is relatively rock-free, making finding meteorites easy. Or easy *enough*. But how do we know when it's a meteorite and not just a regular old rock? (Not that there's anything wrong with regular old rocks.) And, why is Allan Hills so special?

A chat with Barbara Cohen, a planetary scientist at NASA Goddard Space Flight Center, who's worked on multiple NASA missions, was delightfully informative. She's the principal investigator of the Lunar Flashlight mission, a briefcase-sized spacecraft that will use infrared to look for ice on the Moon. She's also an associate project scientist on the Lunar Reconnaissance Orbiter, works on the Sample Return team for the *Perseverance* rover on Mars, and has worked on three other Mars rover missions—Curiosity, Spirit, and Opportunity.

All that work is vitally important for understanding the Moon and Mars, but that was not our topic of conversation. Most germane to this book are her four expeditions to Antarctica to go meteorite hunting, between 2003 and 2017. (She has no current plans to go back, as spaces are often limited and in high demand.)

"Meteorites fall at the same rate all over the Earth, and there's nothing special about any geographic location in terms of how much material is accumulated there," Cohen says. But of what falls, a lot is going to fall into bodies of water, which cover around three-quarters of Earth. Many of them will also fall into remote locations like the Canadian Yukon where no one is going to stumble across them. And according to Cohen, a lot of meteorites that make it to Earth just aren't bound to make it that long.

"When rocks come in from space, they're very unaltered and haven't had a lot of interactions with things like atmosphere and water, and so as soon as they fall onto the Earth, they start to decompose really, really quickly with interaction with their elements," she says. "So if something fell in your backyard, you know, in something like 10 to 20 years, you might not even see it anymore."

While the conception of Antarctica is that it's filled with snow, ice, and penguins, a lot of it is actually a desert—including Allan Hills. Sometimes this just means it's very dry, but other times this means that its ice never really changes because of low (or no) precipitation. So, materials there are exposed to destructive elements like moisture less often. Northwest Africa, especially Morocco, the western Sahara, Algeria, and Mauritania, is also a hotbed for meteorites because of the bone-dry conditions there.

In Antarctica, "the wind is blowing the ice away and so it's leaving a lot of deposits behind of meteorites," while the Saharan sites have similar effects at play. But there are other reasons both sites are ideal for meteorite hunters as opposed to, say, a wheat field in Kansas. "It helps that in both those cases, meteorites are dark against a fairly light background. In the Sahara, it's a tan background, and in Antarctica, of course, it's a white background," Cohen says, adding that there aren't many terrestrial rocks in the mix in many locations. That means that in some cases, the rocks you *do* see are probably not of Earthly origin. We have samples on Earth of at least three solar system worlds—the Moon, Mars, and an asteroid called Vesta— but there are many other samples of undetermined origin or just leftovers from the material that formed the planets, moons, asteroids, dwarf planets, and other assorted solar system bodies.

Each of Cohen's expeditions to Antarctica lasted about six weeks. Some days, Cohen says, they found no meteorites. But on others they found hundreds, collected in sterile Teflon bags and kept frozen until the end of the expedition, at which point they were sent to the Johnson Space Center for cataloging. Cohen says in these areas, you have to adjust to knowing what the terrestrial rocks there look like; many contain feldspar, quartz, or other recognizable metamorphic rocks forged by ancient volcanoes. Finding the meteorite is a matter of, "Same, same, same, same—different . . . oh, what's that?" according to Cohen. Meteorites that survive entry often have a glassy coating. Many of them are black. Some have a fusion crust, which is formed by the melting of the surface layers of the meteorite as it blazes through the atmosphere on the way to the surface, and have a "crazed glaze," she says.

Because meteorites look so different from surrounding rocks, most things that look like meteorites are scooped up in the field and analyzed much later on. "So we don't make any judgments," she says. "We just pick everything up." Some might be smaller than an inch, while a rare few may be larger than a foot and weigh several hundred pounds—although only 30 Antarctic meteorites have been found that weigh more than 39 pounds.[2] The largest meteorite ever found was the Hoba meteorite discovered in 1920 by a farmer in Namibia. It measures around nine feet by nine feet and weighs 66 tons.[3] One way to tell that something might have a planetary origin, instead of being debris, is that it might look like a lava rock, which is more porous than other rocks. (Of course, here, some lava rocks from Earth may sneak their way into the bag . . .) Some iron meteorites are especially heavy too, but weight is never much of a determining factor in the field since meteorites look so alien to everything around them.

Some meteorites that looked remarkable ended up being not that different from others of the same type—they just eroded a little differently and had a different luster. But the opposite could be true, too. On one expedition to the Graves Nunataks region of Antarctica in 2006, Cohen and her compatriots struggled to find meteorites due to bad weather and bad luck. But one day, they happened to find a couple of meteorites that were a color she describes as pistachio green—or so her heavily filtered goggles seemed to indicate. (Please don't eat the meteorites.)* They knew they were *a little* unusual and not ordinary chondrites (the most common type of stony meteorite that includes round molten droplets called chondrules), but they still went in the bag like any other. The meteorites, GRA 06128 and GRA 06129, ended up being the first examples of a new class of achondrites—a type of stony meteorite that doesn't have chondrules—and they weren't pistachio green: they were more of a bright pink under ordinary light in the lab. Both had a high amount of sodium-rich plagioclase feldspar rarely seen in other meteorites and suggests some unusual behavior when lavas cooled on the surface of the meteorites' parent body.[4]

Other places in Antarctica have fragments from the Moon far from any place an Apollo astronaut ventured, helping us piece together lunar history more in depth. (Cohen found a few of these, too, in her expeditions.) But when it comes to bodies like Vesta and Mars, these meteorites are all we have to study in Earth labs right now, making them vital in understanding the history of those bodies from which NASA missions can't yet bring samples back—even though NASA missions have visited both bodies. While no pieces of Venus have made their way to Earth and survived that we know of, pieces of Venus might be on the Moon, according to a 2020 study.[5] Maybe future lunar

---

*Admittedly, geologists are well known for licking rocks, but not for eating them.

expeditions could bring back those pieces as well. And in 2008, meteorites were found in Sudan that had chemical compositions like no other, containing amphiboles (hydrous minerals) absent in other chondritic meteorites and that appeared to be formed in dense conditions—possibly coming from a long-lost solar system dwarf planet.[6]

METEORITE COMPOSITIONS

The Earth continuously accumulates material from space in the form of meteorites, 50 million kilograms (equivalent in mass to about 7,000 elephants) of it per year.[7] Scientists draw these estimates from monitoring how many meteorites enter the atmosphere over a given area as captured by sophisticated cameras, and then applying that figure worldwide. In some ways, we can think of the Earth as still forming itself as it accumulates this material. Although only a small fraction is found by humans as intact meteorites that we can study, the information we learn from them about other bodies in the solar system has greatly helped our understanding of how Earth and all the planets formed.

Virtually all (99.8%) of the meteorites we've found come from the asteroid belt; a handful come from Mars and the Moon. They all can be classified into one of four main types: chondrites, achondrites, irons, and stony irons. Both chondrites and achondrites are made up mainly of silicate minerals, similar to what we find in Earth's crust and mantle, such as olivine and pyroxene. They generally come from the outer crust and mantle-like layers of asteroids that have been broken apart, hurtling material into space eventually to reach Earth's surface.

Chondrites are the most primitive and pristine of all meteorites, by which I mean they formed when small grains of different minerals accreted in the earliest ages of the solar system. The key distinguishing feature of chondrites is that they haven't experienced much heating or chemical alteration since their formation, so we're seeing the tiny grains that formed at the beginning of the solar system in their (almost) preserved state. Basically, they've lived a very sheltered life. It's from these meteorites that we have our best estimates of the age of the solar system (by radioactive dating of the small grains in the meteorite) and also our best estimates of the composition of the material that ended up making planets.

One of the most remarkable figures I've ever seen is a graph with the elemental abundances of all the materials in a chondrite compared with the elemental abundances of material in the Sun's outer layers. Aside from hydrogen and helium, which we know the small planetesimals in the early solar system couldn't attract from the protoplanetary disk, the compositional abundances in these meteorites are remarkably similar to that of the Sun. That is, if you took all the hydrogen and helium out of the Sun and just looked at what was left, the relative amounts of elements like magnesium, silicon, iron, and so on are almost identical to that of chondrites.

There are a few exceptions that we understand, including that the Sun's photosphere has less lithium because it's used up in fusion processes deeper within it, and the fact that the meteorites have less carbon, oxygen, nitrogen, and noble gases because these elements are volatile and would easily vaporize out of the meteorite. We think of the Sun as a ball of hydrogen and helium, but this remarkable finding shows that the original building blocks of planets are identical to those in the Sun: they're made of the same stuff.

Achondrites are still silicate-rich like chondrites, but they come from bodies that have experienced significant melting and some differentiation. They look a lot more like the rocks formed when magma cools on Earth. The minerals in achondrites recrystallize after having been melted into new minerals, and this causes the meteorites to look very different from the pristine chondrites. Their similarity to Earth rocks makes them quite hard to spot visually, unless you're in a place like Antarctica or the Sahara where you aren't expecting any rocks at all.

All the meteorites we have from Mars and the Moon are achondrites. We also have a set of achondrites that come from the asteroid Vesta, as well as from other asteroids and former asteroids in the asteroid belt.

Iron meteorites are mostly made of, well, iron. These meteorites come to us when a differentiated planetesimal (i.e., one that separated into an iron core and rocky mantle) is broken apart by a collision, and then fragments of the core impact Earth. Every time I hold an iron meteorite I think in awe: This thing used to be at the center of an asteroid or planetesimal!

Iron meteorites are one of the major pieces of evidence for the composition of Earth's core. We don't find "copper meteorites" or meteorites made of other metals with similar densities to iron. But we do find lots of iron meteorites. In fact, they're oversampled in Earth's collections, by which I mean that we have way more iron meteorites than we should compared to the other classes of meteorites. This is because it's way easier to identify an iron meteorite as a meteorite since it's so different from most rocks on Earth's surface, whereas you might miss a chondrite or achondrite meteorite as just being another Earth-produced rock* unless you look very closely and know what to look for.

Stony irons look like a combination of achondrites and irons, having roughly equal fractions of iron and silicate minerals. One type of stony iron is called a "pallasite" and it is, without a doubt, the prettiest of all the meteorites. In pallasites, the stony part is in the form of olivine—a magnesium-iron silicate that has a beautiful green color with some outstanding samples called "peridot" used in jewelry. The olivine occurs in large translucent crystals surrounded by a dark

---

*Not that there's anything wrong with Earth rocks.

iron matrix. They too come from former asteroids and plane-tesimals.

One theory of their formation is that they were part of the core-mantle boundary of former asteroids. So we are really see-ing a chunk of the boundary between two layers of a planet. The iron matrix is from the core and the olivine crystals from the mantle. Another possibility is that they might come from melting that occurs during a large asteroid impact onto a body. The melting would separate out the iron in the rock from the silicate minerals, and they would then cool and solidify into a pallasite. Admittedly this is still a cool origin story, but my heart hopes that the core-mantle boundary origin story ends up be-ing the right answer.

DETERMINING A METEORITE'S ORIGIN

Meteorites give us detailed information about the insides of the planets—at least, the rock and metal parts of those insides. Most of the ices and gases present in the planetesimals or as-teroids that eventually impact Earth don't make it into mete-orites because their volatility causes them to vaporize either during their journey to Earth, or their entry through Earth's atmosphere.

But we can sometimes get clues about the volatile com-ponents of meteorite parent bodies hidden inside the meteor-ite. For example, tiny pockets of gas bubbles from Mars's

Heart of Iron: The Psyche Asteroid Mission

You may have seen headlines over the past few years about a "$10 quintillion asteroid" loaded with iron and nickel (and possibly even precious metals and gemstones), the estimated volumes of which could theoretically upset metals markets worldwide. It's an intriguing prospect, but here are a few things to consider about mining 16 Psyche. First, there's the murky legality of claiming ownership of those resources. While the 1967 Outer Space Treaty says that celestial bodies are "free for exploration and use by all States" worldwide, it doesn't cover the economic exploitation of resources or the involvement of private companies (among which several start-ups have come and gone over the years)—a familiar source of conflict over the history of humanity.

Beyond legal and ethical questions, thinking in these commercial terms misses the enormous scientific value that worlds like 16 Psyche provide. 16 Psyche was discovered in 1852, and it's been suspected to be a metal-rich world since at least the 1970s. The presence of iron and nickel and the object's relative density lead some planetary scientists to believe it's a large planetesimal similar to the cores of terrestrial planets.

Benjamin Weiss, chair of the planetary sciences program at the Massachusetts Institute of Technology, is Deputy Principal Investigator of NASA's Psyche mission, which will explore this potato-shaped, 138-mile-diameter world nestled in the asteroid belt. As a planetesimal, 16 Psyche was part of a group of bodies that formed the building blocks of protoplanets, which then formed planets. "Most of them are gone now; they collided to form the planets—16 Psyche's this remnant," Weiss says. "So, in particular, we'll learn about metal worlds or iron-rich worlds among the small body, asteroid planetesimal population, but we also hope not just to understand that population, but in general, to understand how all these other bodies that we see in the universe form."

At one point, 16 Psyche was believed to be the exposed core of a protoplanet, but planetary scientists aren't so sure now—and indeed, Weiss and many other members of the Psyche mission team were among those who proposed that it's something different entirely. Throughout the solar system, 16 Psyche seems to be the only object of its kind. Even if it wasn't a protoplanet once upon a time, it's still likely a large planetesimal somewhat intact from the early solar system, and a unique one at that.

There are hints that 16 Psyche could have some phenomena unseen elsewhere in the solar system, as one of the few worlds truly made of almost nothing but metals. For instance, there are reflections on the surface that could indicate ferrovolcanism. This means that unlike the cryovolcanoes of the icy moons or the magma volcanoes of Earth and other terrestrial bodies, a young 16 Psyche may have had volcanoes that spewed iron and bits of sulfur from a differentiated core and left their mark on the surface, seen in terrain discolored from the surroundings.

The robotic Psyche mission will carry three scientific instruments and use its communications antenna as a de facto fourth instrument, Weiss says. These instruments include:

- a multispectral imager: A camera that will view 16 Psyche with various filters to learn more about its surface, especially topography and areas that are different from the rest, which could confirm ferrovolcanism
- a gamma ray and neutron spectrometer: An instrument that will look at particles bouncing off the surface of the asteroid that reveal its exact chemical composition
- a magnetometer: A device to measure the magnetic properties of 16 Psyche, including investigating the long-frozen magnetic field, as "a record of that could be frozen into the metal rock layers," Weiss says
- radio science: Each NASA mission needs an antenna to radio data home. But the Psyche mission's communications system can also measure the gravity of 16 Psyche. It works a bit like Doppler radar on Earth, waiting to see how a radio signal bounces off of an object to return to an antenna. The mission team can pair the radio instrument with the imager to map the interior structure of 16 Psyche, including corresponding surface features with gravitational quirks below.

16 Psyche may be our only chance to view something akin to the core of a planet up close and learn about the sorts of forces that shaped them. Even if it isn't a protoplanet core, asteroids are much more than a resource to mine. They're a critical story of how planets and moons form, and they can inform us about what may take place around other stars with planets, moons, and asteroids of their own. Although different from these other bodies, 16 Psyche can inform planetary sci-

entists about a class of worlds that seem to be made of almost nothing but iron. While there's nothing as large in the asteroid belt, Mercury is a world that's composed mostly of iron, and there's increasing evidence that iron exoplanets are numerous around other stars, "so there's this increasing recognition that these are kind of a natural endmember [a geologic term for a pure sample] of how planets form," Weiss says. 16 Psyche could tell us more about processes on these distant, otherwise hard-to-study objects.

FURTHER READING

Elkins-Tanton, Linda T., et al., "Distinguishing the Origin of Asteroid (16) Psyche." *Space Science Reviews* 218 (2022): 17. https://doi.org/10.1007/s11214-022-00880-9.

Ganatra, Devanshu, and Neil Modi. "Asteroid Mining and Its Legal Implications." *Journal of Space Law* 40, nos. 1–2 (2015–2016): 81–104.

Rossman, Sean. "NASA Planning Mission to an Asteroid Worth $10,000 Quadrillion." *USA Today,* January 28, 2017. https://www.usatoday.com/story/tech/nation-now/2017/01/18/nasa-planning-mission-asteroid-worth-10000-quadrillion/96709250/.

Shepard, Michael K., et al. "Asteroid 16 Psyche: Shape, Features, and Global Map." *Planetary Science Journal* 2, no. 4 (2021): 125. https://doi.org/10.3847/PSJ/abfdba.

Taylor, R. C., T. Gehrels, and R. C. Capen. "Minor Planets and Related Objects, XXI: Photometry of Eight Asteroids." *Astronomical Journal* 81, no. 9 (1976): 778. https://doi.org/10.1086/111953.

United Nations Office for Outer Space Affairs. "Treaty on Principles Governing the Activities of States in the Exploration and Use of Outer Space, Including the Moon and Other Celestial Bodies." Resolution 2222 (XXI), adopted by the General Assembly, December 19, 1966, signed on January 27, 1967.

Zuber, Maria T., et al. "The Psyche Gravity Investigation." *Space Science Reviews* 218 (2022): 57. https://link.springer.com/article/10.1007/s11214-022-00905-3.

atmosphere are trapped in meteorites from Mars; that's actually one way we know where these meteorites are from. When we compare the atmospheric gases in the meteorites with what landers on Mars have measured, we find they're the same. Since every planetary body has a unique atmosphere, this makes a pretty definitive test of where the meteorite is from.

However, it's hard to do this for meteorites that come from bodies without atmospheres (which is basically all the other meteorites except for the ones from Mars). So how do we determine where they're from? One relatively easy way is if we happen to see the meteor fall to Earth. We can backtrack the incoming trajectory of the meteor to determine where it originated with a network of meteor cameras placed worldwide, and sometimes from amateur footage of meteor entries. However, this method requires that the meteor followed a simple, direct path to us. That's not always the case. Indeed, many asteroids change their orbits because of close gravitational encounters with other asteroids or planets—pushed out of place by the gravity of the other body. This is one reason why projects like NASA's Center for NEO (Near-Earth Objects) Studies are continuously looking for asteroids that might become "potentially hazardous" for Earth. But this method only works if we see the meteorite fall. Most of the over 50,000 meteorites that we have are found in places like Antarctica, long after they've fallen to the ground. We therefore have no record of their incoming trajectory.

Lunar meteorites are relatively easy to identify because we're fortunate to have samples from the Moon brought back by Apollo astronauts. The chemical composition of lunar meteorites is a telltale sign that they came from the Moon.

Another important technique used to identify the original body of meteorites involves looking at their oxygen isotopes. A quick primer on isotopes: An element is defined by the number of protons it has in its nucleus. An element with only one proton in the nucleus is hydrogen, an element with two protons is helium, and so on. Oxygen has eight protons. But along with protons, neutrons can also exist in the nucleus. Neutrons are neutrally charged (unlike the positively charged protons), but they have similar masses as protons. An element can have different numbers of neutrons in its nucleus, and these result in isotopes of that element. For example, hydrogen is most commonly found with one neutron in the nucleus (along with the one proton), but it's also possible to find an isotope of hydrogen with two neutrons (we usually call it deuterium) and with three neutrons (called tritium).

Oxygen has three stable isotopes: ^{16}O (with 8 neutrons), ^{17}O (with 9 neutrons) and ^{18}O (with 10 neutrons). The superscript before the O gives the total number of protons and neutrons in the atom. Nearly all (99.7%) of the oxygen on Earth is in ^{16}O. The remainder is mostly in the other two stable isotopes (there are some unstable isotopes of oxygen, but they decay quickly).

Scientists typically look at the ratios of ^{17}O to ^{16}O and ^{18}O to ^{16}O, and what they find is that all rocks from Earth have the same ratios. All rocks from Mars have a different ratio than the Earth rocks, but the same as each other, and so on. With oxygen isotope ratios, we can determine that a group of meteorites actually came from the same location even if we don't know what that location is. It's also helpful in eliminating whether a meteorite is from Mars, or the Moon, places where we can ground-truth the ratios by comparing them with measurements we've made from instruments on the surfaces.

Incidentally, just as we can tell what planet a meteorite came from by certain isotopes and ratios of chemicals, we can use the same technique to figure out whether something is from outside our solar system. A 2018 study looked at chemicals found in comet dust called "glass with embedded metal and sulfides" (GEMS). Some of the GEMS were encased in a form of carbon that melts away at relatively low temperatures, meaning that they likely formed before the solar system.[8] In addition, a fragment found in a meteorite called Hypatia seems to have come from a distant supernova, determined from how different elements had different elemental ratios than if they'd formed in our solar system.[9] However, both interstellar objects we know of, 'Oumuamua and Borisov, were determined to be from elsewhere based on their trajectory. Telescopes on the ground and in space looked at the movements of the objects, determining that they weren't bound to the sun. If we success-fully snag a sample from a future interstellar visitor, we may be able to learn more about the environment in which it was forged.

Meteorites are the backbone of our understanding of solar system composition. By bringing samples of faraway places to us, they eliminate a major hurdle in gaining the understanding we want of processes inside planets. In some ways, it would be great to have meteorites from Earth's inte-rior in order to understand our own planet better. Imagine finding a pallasite from Earth's core-mantle boundary to help in answering some of our biggest questions about the compo-sition of Earth's core and deep mantle. (This would require the Earth to be broken apart, so in other ways, this is not ideal.)

Luckily, we have something akin to meteorites that bring samples from deep-Earth to the surface while preserving them. And it's all thanks to diamonds.

SAMPLES FROM DEEP EARTH

The saying goes, "Diamonds are a girl's best friend," but my high-pressure mineral physicist colleague tells me the saying should be "Diamonds are a *geologist's* best friend." She men-tioned this at a faculty meeting once, and an enthusiastic dis-cussion ensued regarding whether diamonds or zircons are actually a geologist's best friend.* I'm not sure an agreement was ever reached on the subject.

Diamonds form in the deepest and oldest parts of continental crust—regions called cratons—as well as the upper mantle at depths ranging from 100 to 400 miles, where the pressure and temperature conditions are just right to transform carbon into the diamond phase. They get close to the surface when they get entrained in volcanic pipes called kimberlites and hop a ride in magma to get closer to the surface.

Aside from their use in jewelry, diamonds in the Earth are very helpful in understanding the interior of the planet for two reasons. First, although pure diamonds are considered most valuable in the jewelry world, diamonds with impurities or "inclusions" (i.e., little pockets of fluids or other minerals that are enclosed inside diamond crystals) are most beloved by geologists. Admittedly some jewelers like inclusions as well; it's what can give color to diamonds and the reasons there are such things as yellow, green, pink, and of course my favorite: "chocolate" diamonds.

But geologists love these impurities because they bring vital information about processes in Earth's interior to the surface. The fluids or minerals are encased in the diamonds, where they don't interact with anything else. The diamonds are like

---

*Zircons are highly durable, changing little from the time they were formed. Often uranium is trapped within them, and its rate of decay allows geologists to date a rock sample.

little preservation centers that we can use to study what the composition and conditions of Earth were like at the time the inclusion was preserved. This also means that diamond inclusions can be used as time capsules. For example, diamond inclusions have been used to help determine when the Wilson cycle of plate tectonics started on Earth (although there is still debate on how the age of the zircons relates to when plate tectonics started),[10] and they provide evidence that water exists at the boundary between the upper and lower mantle.[11]

Recently, a group of scientists from Arizona State University, the University of Chicago, and the University of Hawaii at Manoa found evidence that diamonds may even exist much deeper in the Earth, near the core-mantle boundary.[12] This isn't because they collected samples of these diamonds, but because they made them after replicating compositions and conditions similar to that of the core-mantle boundary. Interestingly, in order to make these samples, you actually need to use instruments made of diamonds themselves.

MAKING SAMPLES

Diamonds can be used as a tool to compress materials to really high pressures in a device known as a diamond anvil cell. Most people are familiar with the fact that diamond is a really hard substance (i.e., it requires extreme forces to deform). This can be exploited in squeezing materials.

Let's say you have a sample of some rock or mineral that you want to study at pressures similar to those deep inside the Earth. You can place that sample in between the pointy tips of two diamonds and then force the diamonds closer together by squeezing them in a press. This works because you can achieve a high pressure by taking a large force and concentrating it over a tiny area, like the tip of a diamond. Because diamonds are so hard, they don't break under the pressure like most other materials do. So the force we apply to the fatter side of the diamond gets amplified when it gets to the tip and squeezes the sample to high pressures.

We can then study what happens to the sample and ask questions like: Does it melt? What's its density? What's its electrical conductivity? Does it change crystal structure? Do any chemical reactions between different compositions occur? It's with experiments like this that we can learn about the properties of the materials deep inside planets since we can't go get samples of them directly.

So how did the scientists make diamonds with a diamond anvil cell? They started by considering the form of carbon that would likely exist deep inside Earth's core—the molecule Fe_3C (that's a molecule with three iron atoms and one carbon atom). Since the previous studies discussed above had shown that water exists in Earth's deep mantle, the scientists considered the possibility that this water also exists near the core-mantle boundary. (Knowing where all the water is in Earth is important to fully understand the water cycle—how water moves between the atmosphere and Earth.) The scientists then asked: What happens when Fe_3C interacts with water at one million bars of pressure and several thousand degrees in temperature (i.e., the conditions that these materials would experience at the core-mantle boundary)? Their experiments showed that the Fe_3C reacted with the water to form pure diamond along with other compounds.

Clearly it's cool to think that there are diamonds at the core-mantle boundary, but this result may also explain a mystery related to Earth's carbon cycle: there seems to be too much carbon in the mantle. That is, if you consider Earth's building blocks and you differentiate the planet into a mantle and core, most of the carbon should have gone to the core, leaving only a tiny residual in the mantle (1–5 parts per million).[13] But in reality, evidence suggests there's about 20–100 times more carbon than that in the mantle. This discrepancy could possibly be explained by interactions like the ones observed by these researchers, where the carbon in the mantle comes from reactions with the core.

This carbon in the mantle is a vital part of Earth's carbon cycle, responsible for everything from the formation of limestone rock to the bones in our body. The amount of carbon in Earth's atmosphere is also partly determined by how much carbon is released from Earth's interior, whether by humanity's burning of fossil fuels or by the eruption of carbon in the form of carbon dioxide from volcanoes.

Samples from other planets and from Earth's deep interior have proven to be a vital resource for learning about planetary interiors. But if this were the only source of information we had on planetary interiors, our knowledge would be quite limited. If we can't directly analyze or see parts of a planet we want to know about, we'll need other, more indirect methods to get more information. Luckily, scientists have designed the right tools for this.

CHAPTER 4
Fierce and Formative Forces

THE ICONIC PHRASE "USE THE FORCE!" from Star Wars movies translates easily in real life to studying the interior of planets as it does to the fictional energy field in the movies, where the forces both experienced by and *caused* by planets provide copious information about what's happening inside these enormous spheres. The previous chapter showed us how samples from the planets can provide information on their interiors; these forces provide complementary information that gives us a more holistic picture. And they're part of what makes life on Earth possible.

The forces we're talking about are our old friends from high school physics classes: gravity and electromagnetism, along with seismic forces, which really are just the manifestation of pressure and tension forces. We'll consider gravity and electromagnetism in this chapter and discuss seismic forces in the next.

GRAVITY'S RELATIONSHIP TO MASS

As a Gen-Xer, a mainstay of my TV-watching life in high school and college was the sitcom *Friends*. In one of my favorite scenes, paleontologist Ross is shocked to find out that his quirky friend Phoebe doesn't believe in evolution. At some point, exasperated, he says something to the effect of "That's like saying you don't believe in gravity!" to which she responds that she's not sure about gravity, either. As Ross grows more frustrated, asking what she possibly can't believe about gravity, she states, "Lately I get the feeling that I'm not so much getting pulled down as I'm being pushed."

Although I love both the visual and the idea of something coming along and pushing us toward Earth to keep us from floating away into space, in reality, it's more accurately described as a pull. Gravity is a force of attraction between two objects with mass. Admittedly, the "pull" of gravity can act in unexpected ways. For example, objects in orbit around the Earth, like the International Space Station, communication satellites, or even our Moon, are really connected to the Earth because of gravity's pull, even if they never actually reach it. They're inextricably bound to us with an invisible tether.

Because of the relationship between gravity and mass, measurements of gravity are key to determining the most fundamental property of a planet's interior: its density. If we know the mass and volume of a planet, we can deduce its density, from which we can gather information about what the planet is made of.

Graduate students in my department at Hopkins must take a qualifying oral examination that helps us assess their knowledge base and ability to apply scientific reasoning—important components of attaining a PhD. Three professors sit across a table from a student and ask questions. One of my favorites is: "How do you weigh a planet?"

What I really mean by this question is: How do we know the mass of each planet (and hence, how do we know what they're made of)? Go to Wikipedia and you'll find masses quoted for each planet. They're staggering numbers. Earth's mass is about 6×10^{24} kilograms. That's a six followed by 24 zeros. I can't even put that into a visualized unit like a number of elephants because there would still be an obscene number of zeros—the number is so huge, there's really nothing reasonable to compare it to. Indeed, when planetary scientists quote other planets' masses, we sometimes use Earth's mass as a "unit." For example, Jupiter's mass equals 318 Earth masses.

So back to the question, how do we weigh a planet? (And yes, I know there is a difference between mass and weight, but let's be colloquial here.) We obviously don't have scales on which to stick a planet, even if we figured out how to get that scale in the right place. The answer lies in using the power of gravity.

In 1618, Johannes Kepler took the first empirical steps to figuring out the key reasoning needed here in his third law of planetary motion: "The squares of the orbital periods of the planets are directly proportional to the cubes of the semi-major axes of their orbits." (The semi-major axis is the long axis of the ellipse of a planet's orbit.) This definition is a mouthful, but Kepler was expressing a relationship he noticed for all the planets in the solar system known at that time. When Uranus and Neptune were discovered later, they also followed this relationship, providing further confirmation of the accuracy of Kepler's third law.

Ultimately, Kepler's guiding orbital principles are telling us that the length of a planet's year (i.e., how long it takes to make one complete orbit around the Sun) is related to the distance of the planet from the Sun. This law guarantees us that the closest planet has the shortest year—Mercury's year is 88 Earth days—and the farthest planet has the longest year—Neptune's year is 60,190 Earth days, or 165 Earth years.

In order to understand *why* this relationship holds, we had to wait about fifty years for Isaac Newton to develop his law of gravity in 1686 and laws of motion in 1687. The key was in the "proportional to" bit of Kepler's third law. The square of the orbital periods is actually equal to a number times the mass of the body being orbited times the cube of the semi-major axis. For those who want a peek at astrophysics mathematics, here's how this looks in an equation:

$$(semimajor\ axis)^3 = 1.7 \times 10^{-12} \times mass \times (period)^2$$

If we rearrange this equation, we get an equation for the mass of the body that's being orbited:

$$mass = 5.9 \times 10^{11} \times \frac{(semimajor\ axis)^3}{(period)^2}$$

This shows that for the planets orbiting the Sun, we can actually determine the mass of the Sun by solving this equation using the observed orbital periods and distances of the planets from the Sun.

What about determining the masses of the planets? The beauty of Kepler's third law is that it holds for any object in orbit, which means that it holds for moons orbiting planets as well (in this case, the mass in the equation above is the mass of the planet being orbited). For example, we could actually determine the mass of Earth by using the Moon's orbital period and distance from us. (There are easier ways to determine Earth's mass, though, since we live on the planet and can perform other experiments to measure Earth's gravitational acceleration right here on the surface.) With this equation, we could determine the masses of Mars, Jupiter, Saturn, Uranus, and Neptune by studying the motion of these planets' moons. And we can do it all from the comfort of our own homes (assuming we have an appropriate telescope). In fact, the mass of Pluto wasn't known until Pluto's moon Charon was discovered in 1978.

EXTRAPOLATING THE MASSES OF MOONLESS WORLDS

This all sounds cut and dried until we realize that Mercury and Venus don't have moons. That means we can't use this method to determine their masses. Luckily, you don't have to be orbiting a planet to experience its gravitational force.

Another way to measure a planet's mass is to measure its gravitational acceleration, its "g," which is directly related to the mass of a body toward which something is accelerating. All we need to measure the mass of Mercury and Venus is something to be close enough to these bodies that they feel the pull of gravity and we can measure their accelerations.

For Mercury and Venus, that was first accomplished using asteroids that roam the inner solar system. But that requires us to find and observe these asteroids at the right time and place. It becomes a lot easier to measure the gravitational acceleration when we send spacecraft to the planet. Then the spacecraft itself feels the gravity while approaching a planet, and we can use that information to determine its mass. And if we send spacecraft near moons of planets, then we can use the same technique to determine the masses of the moons themselves.

It's from these mass determinations, along with volume, that we can determine a planet's density. From this, we know the terrestrial planets are made mostly of rocky materials, whereas the gas giant planets are composed mainly of hydrogen and helium, and the ice giant planets are made of materials that were in an icy state when the planets were formed—materials like water, ammonia, or methane. That's because the densities of those materials match the densities measured for those planets.

EXQUISITE DETAILS FROM GRAVITY

Our use of gravity doesn't stop at determining the density of a planet. On the surface of the Earth, we typically say Earth's gravitational acceleration is 9.81 meters per second squared (m/s^2). But that's really an average. If we were to walk to different places on the Earth with sensitive instruments to measure gravity (called gravimeters), we would see that the gravity varies from place to place. This variation is less than 0.1% of the average 9.81 m/s^2 value we use in our daily lives (or at least, in intro physics and engineering classes).*

There are two causes of these variations, but they both relate to one common fact: the gravity experienced at a location on Earth is determined by the amount and *distribution* of mass below that location. If you were standing at a spot on Earth's surface above dense material like an ore deposit, the gravity would be higher there than if you were standing over less dense

---

*Astronomers, being astronomers, just use 10 m/s^2.

material, like an underground aquifer. So if the Earth were weird enough to be made up of columns from the center of the Earth to the surface of different materials—say, iron in one column, carbon in another column and water in a third column—you would experience different gravitational accelerations while you were standing over each of the columns. The gravity would be highest over the iron column and lowest over the water column. From a practical life-on-Earth standpoint, such gravity studies are used by exploration companies as a kind of treasure map of the interior to search for resources like oil and gas.

Although Earth is not this weird column planet, we do measure differences in Earth's gravity due to smaller differences in density under specific locations. For example, subduction zones on Earth, like the Marianas Trench in the western Pacific Ocean, have old, cold, and hence dense lithosphere (the outer brittle layer of Earth broken into plates) descending back into the Earth. These subducting slabs are slightly denser than the material around them. That means the gravity over the slabs is higher. You can actually see where all the subduction zones are on Earth by looking at maps of Earth's gravity field made by spacecraft like the *GRACE* orbiters, which were in operation around Earth from 2002 to 2017.[1]

Another reason for gravity variations on Earth's surface is that Earth isn't a perfect sphere, but, instead, varies in altitude along the surface. There are some big variations, like the tidal

The Gravity Recovery and Climate Experiment

The GRACE mission (2002–2017) featured twin satellites that followed each other in close position in order to precisely measure Earth's gravity field. They used microwave ranging between the two satellites to accurately measure the changes in speed along their orbits. These changes in orbital speed are used to determine gravity differences in the Earth.

By studying changes in gravity over time, GRACE provided crucial data on important climate indicators like the thinning of ice sheets (see figure below), melting of glaciers, changes in deep ocean currents, and the flow of water in and out of underground aquifers.

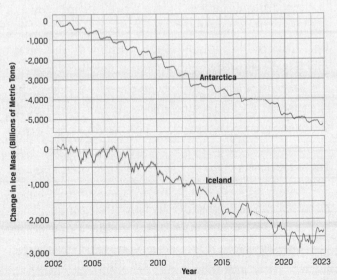

Data source: Ice mass measurement by NASA's GRACE satellites.

bulge that makes Earth football shaped and the rotational bulge due to Earth's rotation. But there are also smaller changes in altitude caused by, for example, mountains. When you're standing atop Mount Everest, there is more mass between you and the center of the Earth—and thus, more gravitational force—than when you're standing at sea level in Baltimore.

However, it turns out there are two subtleties that can make it less clear whether the gravity at the top of a mountain will be higher or lower than at sea level. First, the higher you are up a mountain, the farther you are from the deep interior of the Earth, where the density is highest. Gravity really cares about how far you are from the mass, and being further away means you are less attracted to that central mass.

Second, mountains aren't just plopped onto the surface of the Earth. The lithosphere floats on the asthenosphere (the weak, deformable upper mantle layer) below it. That means that mountains produce lithospheric roots that plunge into the asthenosphere, just like icebergs plunge below the surface of the water. These lithospheric roots are less dense than the surrounding material, and so it's possible for the gravity on mountains to be lower than at sea level. The devil here really lies in the details of the mountain and the surroundings, and that's good news, because it means we can use these gravity measurements to map the density inside the Earth.

We also have gravity maps for other planets for which we have orbiting spacecraft. For example, the most stunning features in the Moon's gravity map, produced by the GRAIL mission, are the plethora of circular features encompassing the body.[2] These are all impact craters! Giant impact craters like Hellas and Argyre are also prominent in the gravity field of Mars, along with Tharsis Mons, a giant volcanic province. These were produced from the Mars Global Surveyor, Mars Odyssey, and Mars Reconnaissance Orbiter missions.

GRAVITY AND TIDES

So far we've focused on how gravity can be used as a tool to study a planet's interior; but gravity is also a major force in shaping a planet's interior. First, there's the fact that planets are held together in a spherical shape because of the nature of gravity and the fact that the gravitational force is directly related to the distance between masses. For a homogeneous sphere, that means that every point on a spherical surface feels the same gravitational force from the sphere of mass under it. Without this self-gravity, planets would never have formed in the first place since they couldn't hold themselves together.

But planets don't just experience self-gravity. When a planet orbits the Sun, it also experiences the Sun's gravity. One consequence of that is the planet's orbital motion. But there is another consequence because different parts of a planet will be different distances from the Sun at any given time.

Think of a location on Earth at high noon versus the antip-odal (i.e., farthest) point, which is experiencing midnight. For example, the location on the Earth directly opposite Baltimore is just off the western coast of Australia. The closest city is Augusta, Australia. When it's high noon in Baltimore, it's midnight in Augusta, and Augusta is one Earth diameter farther from the Sun than Baltimore. That means that the gravitational pull of the Sun on Baltimore is slightly larger than the pull on Augusta. This causes the Earth to deform from a sphere into a very slightly elongated (and not easily perceivable) ellipsoid with the long axis between Baltimore and Augusta. As Earth rotates through its day and different locations reach high noon and midnight, this ellipsoid long-axis moves around the Earth, creating what is known as the solar tide.

It's hard to imagine that Earth's rocky surface is being continuously pulled apart like this. The exact same thing is happening to Earth's bodies of water and creates the better-known water tides. But in reality, the rocky parts of Earth are also deforming, albeit on a much smaller scale (inches instead of feet), because rocks are better at holding their shapes than liquids are.

There's a similar tide caused on Earth by the Moon since Earth also feels the Moon's gravity. This creates the lunar tides. That means Earth's shape is kind of weird, with longer axes facing both the Sun and the Moon. Although the Moon is much less massive than the Sun, it's much closer to the Earth. For gravity, that means the lunar tides are about twice as high as the solar tides.

Earth's surface and interior are continuously deforming due to lunar and solar tides. The magnitude of the tidal stretches is directly related to the physical properties of the material inside Earth. We can actually use the magnitudes of the tides to determine the elastic and viscous properties of Earth as a whole, and this constrains what Earth is made of. For example, if Earth's iron core were twice as large as it really is, the tides would be much smaller. And if Earth were entirely made of water, the tides would be much larger.

Tides not only change the shape of the Earth and other planets; they can also create heat inside a planet. That's because when a planet flexes, the rocks making up its mantle and core constantly rub against each other, causing frictional heat. It's the same effect you get if you bend a wire or credit card over and over again. The deformed spot becomes warm. This tidal warming isn't a significant source of heat in Earth because the tides aren't too big. However, it can have spectacular consequences for other planetary bodies, including volcanic activity—both the fiery molten form we have on Earth, and the ice-spewing variety on Enceladus and Europa (for more on volcano mechanics, see chapter 6).

My favorite example is Jupiter's moon Io. Using his famous telescope, Galileo discovered Io in 1610 along with the moons Europa, Ganymede, and Callisto—known together as the

Galilean satellites of Jupiter. Io is the closest of these satellites to Jupiter and hence feels the strongest tidal forces. But there's a complicating factor that ends up happening a lot with moons: Io is gravitationally locked to Jupiter, always showing the same face to it, and hence the tidal bulge is fixed along one axis. Essentially, there's a permanent high noon and midnight on Io with respect to Jupiter. This means Io shouldn't be heated by a continuously shifting tidal bulge in the moon.

But another factor comes into play. Io's orbit around Jupiter is influenced by the nearby moons Europa and Ganymede. This causes Io's orbit to become more elliptical. The elongation of Io's orbit means that it is at different distances from Jupiter at different times along its orbit. This change in distance constantly changes the gravitational force it experiences from Jupiter and hence the amplitude of the tidal bulge. Through measurements of Io's gravity field and amount of tidal heating, we know Io's tides can reach 300 feet, and this change in shape is so strong that it heats Io's interior enough to melt it.

One of the greatest twentieth-century predictions in planetary science was about Io. The Voyager 1 mission, which visited Jupiter and Saturn, was scheduled to fly by Io on March 5, 1979, with the aim of returning our first resolved images of the moon. Three days before the encounter, scientist Stan Peale and colleagues published a paper in *Science* magazine predicting that Io's tidal heating would mean that the moon would be mostly molten and covered in volcanoes.[3] Days later, Peale's prediction was proved true when we received images of active volcanoes covering Io's surface, spewing lava hundreds of miles above the surface before landing back down. This was a thrilling moment of a theory becoming empirical fact; it's moments like these that inspire scientists worldwide.

We've had more opportunities to observe Io since Voyager 1. The Galileo mission to the Jupiter system in the late 1990s and early 2000s provided spectacular images and data on Io. We caught volcanoes in the act of erupting and lava constantly changing the face of the moon. We now know that the tidal heating experienced by Io makes part of the interior molten; essentially this moon has a giant magma ocean beneath a surface riddled with volcanoes.

MAGNETIC FIELDS

Gravity is a powerful means to investigate a planet's interior because the gravitational force acts indirectly. Essentially, we don't have to be in contact with the mass to feel the gravity from it. The mass creates a gravitational field in all of space. This allows us to probe places we can't physically touch or see, like a planet's insides. The same is true for the magnetic force.

As discussed in chapter 1, many planets have magnetic fields generated by a dynamo (the process where the energy of motion is converted to magnetic energy) in their deep interiors. Those magnetic fields extend outward beyond the planet, where we can detect them. By looking at the properties of these magnetic fields and how they change, we can learn about what's happening deep inside a planet. In particular, an active dynamo tells us that a planet must have an electrically conducting fluid in its interior and that complex motions are occurring in that fluid in order to generate the dynamo. Here are some of my favorite examples of how we've used magnetic fields to learn about planetary interiors.

The Unexpected Dynamo at Mercury

The fact that Earth has a dynamo tells us that part of the iron core (the electrical conductor) must be liquid. Although seismology research gave us this information about a hundred years ago, the fact that the core is convecting, producing the churning motions that invigorate the dynamo, is best discerned by the presence of Earth's magnetic field.

The presence of liquid layers in other planets is quite challenging to discover without seismology. There are techniques that involve studying the wobbling of the planet's spin, as detailed in the next chapter, but these require careful measurements over a long period. The detection of a dynamo, however—for example, by a spacecraft carrying a magnetometer—is relatively easy,* and it can change our understanding of a planet immensely.

---

*Admittedly, "relatively" is doing a lot of work in this sentence since flying a magnetometer on a spacecraft to another planet is never "easy."

Take Mercury, for example. The smallest and innermost planet in our solar system is a great example of why you shouldn't judge a book by its cover. At a quick glance, Mercury's gray, crater-filled surface makes it hard to distinguish from Earth's Moon. The Mariner 10 mission in the mid-1970s was sent to investigate Mercury and provided us with the first detailed information of its surface and interior. As such a small body, Mercury was assumed to have cooled down to the point where its interior would be solid. Predictions were therefore made before the arrival of *Mariner 10* at Mercury that the planet would not have a dynamo.

But then an intrinsic magnetic field measured by *Mariner 10* challenged the original assumption. Planetary scientists went back to their drawing boards to determine how Mercury— such a small, relatively cool planet—could have maintained a liquid core. We're still trying to understand it, fifty years later. The most plausible answer we have is related to the composition of the core. If Mercury's core were pure iron, then it would be solid at the relevant temperatures. But what if a small amount of sulfur were mixed in with the iron? Sulfur acts like an antifreeze, allowing the iron-sulfur mixture to be liquid at temperatures where pure iron would be solid. It's the same principle behind salting roads during freezing temperatures to keep water liquid rather than turning into solid ice.

It may seem arbitrary to mix in an element like sulfur to the iron, but geochemists who study how the building blocks of planets form and alter in different environments, have found that when a planet the size of Mercury is forming, sulfur would have readily mixed into the iron and descended to the core during the differentiation stage of planet formation. So it's not too surprising that there is a bit of sulfur in Mercury's core. Indeed, there are probably other elements as well, like a bit of silicon or oxygen, and geochemists study what the implications for having these compositions in the deep interior could mean for the planet.

While the initial assumption that Mercury shouldn't have a dynamo turned out to be false when we actually went to look for it, it was a completely reasonable idea, based on the information scientists had at the time. But this goes to show how important empirical data is in testing theories and determining the truth—in planetary science, and in everything else.

Life on the Lithosphere

Earth's magnetic field played a vital role in discovering that Earth has plate tectonics—the process by which the outermost layer of Earth, the lithosphere, is broken up into plates that move relative to each other on the surface; they are continuously created at mid-ocean ridges and descend back into Earth at the subduction zones.

In the mid-twentieth century, Earth scientists were busy piecing together a puzzle of how Earth's surface works by synthesizing clues found on Earth's surface, such as the alignment of the eastern edge of the South American continent with the western edge of the African continent, the appearance of similar mountain ranges and fossils in distant locations, and the discovery of tropical fossils at high latitudes and marine fossils at the tops of mountains. These clues led scientists to propose the theory of continental drift: that the continents could move over the surface of the Earth.

This theory was first proposed in an empirical way by German meteorologist Alfred Wegener in 1912, but it was vigorously rejected by geologists at the time. When I'm asked why, I like to joke that people wouldn't believe a theory from someone who's so often wrong about predicting the weather. But in reality, continental drift theory was rejected because of an untenable requirement: that the continents move over the surface of the ocean floor. Anyone who has had to drag a filing cabinet on a carpet can understand the issue: You need a lot of force to overcome the friction of the carpet, and with the size of the continents, any mysterious force large enough to do so would likely cause the continents to break up into tiny pieces over the rough ocean floor.

It turns out that Wegener (who also introduced the collective name Pangea for the pre-drift earthform) was on the right track but was missing some important information. That information came with the renewed exploration of the oceans that took place after World War II. For instance, in 1952, Marie Tharp discovered, through sonar, a giant mountain range at the bottom of the Atlantic Ocean, running along the center of the ocean starting north of Iceland and terminating at a latitude near the tip of South America. Called the mid-Atlantic ridge, it's about 10,000 miles long. Some portions of the ridge are actually above sea-level, forming islands like Iceland and the Azores. Sonar also enabled thorough mapping of deep trenches in the Pacific Ocean, and in 1960, Jacques Piccard and Don Walsh embarked in a craft called the *Trieste* to the Challenger Deep portion of the Marianas Trench. In 1960, Harry Hess famously noted the ocean floor's uniform composition, young age, and lack of sediment—and eventually determined that this ocean floor eventually descended back into the Earth at oceanic trenches. How could the continents be so much older than the ocean floor, especially if the continents were supposed to move over it?

The answer came from studies of the magnetism in the rocks that made up the ocean floor. If one were to stand at the volcanic source along the spine of the mid-Atlantic ridge (not recommended) and measure the magnetic field on either side of the ridge, you'd find they're about the same. As you move further away from the ridge on either side, you'd find that the magnetic fields in the rocks were the same at the same distances from the ridge.

The mid-Atlantic ridge is effectively surrounded by bands of identical crustal magnetic fields on either side. This crust was magnetized when it cooled down from its initial lava state and froze in the magnetic field at the time of its formation. For these bands of rock to have identical magnetic fields would mean that the bands on either side of the mid-Atlantic ridge would have to have formed at the same time. The only feasible explanation for this was that each band of rock initially formed at the mid-Atlantic ridge, then spread out away from the ridge on either side. Then new rock would form from the volcanoes at the ridge and spread out on both sides as well. So the ocean floor was constantly being created at the mid-Atlantic ridge, then moving away from it on both sides.

The mechanism of plate tectonics was then better understood: it wasn't that the continents were drifting over the ocean floor, but rather that the entire surface of the Earth, including the ocean floor, was drifting around. The combined oceanic and continental lithosphere is broken up into plates that move relative to each other. This fixed the problem Wegener had, and geologists were finally convinced. A vast range of data collected since these initial measurements has continued to confirm that plate tectonics is the process by which much of Earth's surface evolves. And plate tectonics is just the surface manifestation of the convective motion happening throughout the entire mantle in Earth, effectively cooling the planet down and responsible for everything from Earth's carbon cycle to the generation of Earth's dynamo. And magnetic fields were a key piece of the puzzle.

OTHER PLANETS' PLATE TECTONICS?

There is no strong evidence of plate tectonics currently acting on other rocky planets, although scientists have hypothesized that Mars had plate tectonics early in its history.[4] We don't see evidence of subduction zones and ridges on other planets today; we believe Earth is unique in the solar system in this aspect. The other planets experience a form of mantle convection known as stagnant lid convection. The stagnant lid is the lithosphere, and it just stays on the surface. But at some depth below, the mantle rocks are convecting.

We don't fully understand why Earth is the only planet we know of with plate tectonics. The leading theory involves recognizing two unique features of Earth's interior that make plate tectonics work: First, the oceanic lithosphere of Earth is relatively thin compared to the lithospheres of other planets. Mars's lithosphere, for example, is about 300 miles thick, which is about 10% of Mars's radius. Compare that to Earth, where the lithosphere is only about 40–100 miles thick, which is about 1% of Earth's radius.

The thinness of Earth's lithosphere may be directly related to the fact that Earth is the largest of the rocky planets. Smaller planets cool faster, and the cooler the planet, the thicker the lithosphere. When the lithosphere is thick, it's hard to break it up, which is what's needed to get the lithosphere to subduct back into the Earth.

The other important factor is that the upper part of the Earth's mantle, called the asthenosphere, is about a hundred times less viscous than the lithosphere. That means that it's relatively easy for the lithosphere to move around on top of the asthenosphere, resulting in plates that are mobile enough to allow spreading at mid-ocean ridges and subduction. I compare it to a crème brulée. After you crack the crust, it doesn't take much effort to move the caramelized sugar surface over the custard filling.

We think the main reason the asthenosphere is so much less viscous than the lithosphere is because it contains a bit of melted rock (less than 0.1%) that acts as a sort of lubricant, allowing rocks to slide past each other more easily. It's in the asthenosphere that the pressures are low enough that the high temperatures of the rocks put them close to their melting points. There is no evidence for an asthenosphere on other rocky planets, and again, this might be due to their relative sizes.

Plate tectonics plays a major role in making Earth a habitable place. It's involved in the carbon cycle and the recycling of other volatiles, responsible for our breathable atmosphere and liquid water on the surface. Plate tectonics is also the fastest way to cool a planet since the hot rock in the deep interior comes right to the cold surface. If Earth didn't have plate tectonics, and instead had a stagnant lid, it's possible that the Earth wouldn't cool fast enough for the iron core to convect and produce our magnetic field. Plate tectonics may therefore also be indirectly responsible for shielding us from the harmful radiation of the solar wind.

A MISUSE OF MAGNETIC FIELDS: SATURN'S LENGTH OF DAY

In 2004, the Cassini mission was approaching the Saturn system, ready to shed light on some of the secrets of the glorious ringed planet. I was a postdoctoral fellow at MIT at the time, and I was giddy about the latest results I had just read.[5] I showed up to a group meeting and (not so quietly) proclaimed, "Good news! Saturn's rotation period has gotten six minutes longer!" The rotation period is the length of the day, and I was referring to the fact that Cassini's measurement of Saturn's day-length was different from that measured previously by the *Voyager* 1 and 2 spacecrafts in the early 1980s. A very distinguished professor in the department frowned and said, "A planet can't change its rotation period that much in 20 years, it's impossible!"—to which I replied, "Exactly: that's why it's good news!"

When a seemingly impossible thing happens in science, it's usually because of a mistaken assumption made in coming up with a conclusion. The same was true in this case. It absolutely would have been impossible for Saturn to change its rotation period that much because the planet is confined to follow the law of conservation of angular momentum, which states that an external torque is needed to change a planet's rotation rate. Planets can't just stop spinning, or change their spin speed, without some major external torque acting on it, or changing its shape (recall the figure skating analogy in chapter 2 when we discussed how the solar system formed). Saturn didn't have either. But its rotation period was longer . . . or was it?

Here's where the assumption came in. It turns out that measuring the rotation period of a giant planet isn't so straight-forward. For the rocky planets, it's (relatively) easy: You mark a particular location on the surface (like a crater, or volcano, basically something that doesn't move), wait for the planet to spin around and for that location to come back to the same point, and that gives you the time for one full rotation. But the giant planets don't have surfaces. What we see is their atmo-spheric layers, and there are winds in those layers. We know the winds in Earth's atmosphere don't move at the rotation period of Earth and that they're variable in time. Getting a precise measurement on the length of Earth's day by measuring the winds would not be smart.

But why is it important to know a planet's rotation rate? As far as we know, there aren't aliens with watches marking off the hours of a Saturnian day who might care if it's a few minutes longer or shorter. Precisely knowing the day length for other planets is important for us here on Earth because a planet's ro-tation causes it to bulge at the equator. As we'll see in the next chapter, the size of the bulge is related to the mass distribution inside the planet; we can learn about the inside of the planet from the shape of the bulge.

So, how do we measure the rotation rate for giant planets? Although the winds we see in their atmospheres move around a lot, these planets' denser deep interiors are more steady. If we could see some feature deep inside the planets, we could use that as our fixed point to determine the length of the day. But, alas, all those layers of atmosphere block us from seeing the deep interior.

Planetary scientists get around this nuisance by using a giant planet's magnetic field. The magnetic field is rooted deeper in the planet, below the atmosphere. And although the magnetic field can change in time, it does so on much longer time scales than the rotation period of the planet. So scientists looked for a feature caused by the magnetic field and waited for that feature to reappear as the planet completed one rotation.

The feature they focused on was actually the radio emissions that radiate from the planet at its magnetic poles. These radio emissions occur because high-energy particles in a planet's magnetosphere spiral along magnetic field lines and bounce at the magnetic poles—the same process that causes auroras on Earth and some of the other planets.

One important fact needed for this method to work to determine the rotation rate is that the magnetic pole has to be offset from the geographic pole. Take Earth as an example. Earth's magnetic pole is about ten degrees in latitude away from the geographic pole. If we were in a spaceship watching Earth rotate below us, the magnetic pole would directly face us only at one instant in the planet's rotation. But if we could record the radio emissions beamed at us when the magnetic pole is directly below us, we could look at the time between receiving these radio emission signals and use that to determine Earth's rotation period. Of course, there are easier ways to measure the rotation period for the rocky planets, but the rotation periods of the giant planets are all determined from this technique.

And it generally works. But Saturn proved to be the exception, and in hindsight, it's not that surprising that this technique was doomed to fail for Saturn. That's because Saturn's magnetic field is somewhat unusual for a planetary magnetic field. Our first measurements of the field by the Voyager 1 and 2 missions in the early 1980s found that the field has incredible symmetry about its rotation axis. This "axisymmetry" essentially means that if you were to trace the magnetic field along any line of latitude around the planet, the field would be exactly the same all along that latitude. So you essentially couldn't tell what longitude you were at based on the magnetic field. Another way to think of this is that the magnetic pole at Saturn is at exactly the same point as the geographic pole.

You might be thinking: But if Saturn's magnetic field is axisymmetric, how could we use the technique of detecting the period of radio emissions in order to determine the length of day? You'd be right—in principle, we can't. And yet there was this repeating feature that could be seen in radio emissions coming from Saturn. After the Voyager 1 and 2 missions, scientists believed they needed to measure Saturn's magnetic field more accurately, and that when they did so they'd realize the magnetic and geographic poles aren't completely aligned and we would be okay in explaining the periodic radio emissions.

Enter the Cassini mission hurtling toward Saturn in the early 2000s. It too detected these radio emissions, and it was the period of these radio emissions that had become six minutes longer than what was detected from Voyager. So how could that be? Realizing that it was impossible for Saturn's rotation period (and internal magnetic field) to change that much over 20 years, scientists looked to the other half of the equation: the atmosphere and magnetosphere that supply the high-energy particles that end up beaming the radio emissions.

Scientists ultimately learned that a previously undetected giant polar twin-vortex within Saturn's uppermost atmosphere has swirling winds with speeds of more than 4,000 miles per hour—like some science fiction hurricane. These churning winds were perturbing the magnetic fields emanating from the interior of the planet and creating the observed radio emissions. It was changes in this twin vortex that caused changes in the measured period of the radio emissions.[6]

But this led to a new problem: If the period that scientists measured was caused by the atmosphere perturbing the magnetic field, then how can we measure the rotation period of Saturn? By the end of the Cassini mission in 2017, we had also obtained much more accurate and detailed measurements of Saturn's magnetic field. Those didn't help with the problem, because the field was determined to still be completely axisymmetric, even with the more complete data. We weren't going to be able to use the magnetic field to accurately determine the rotation period of Saturn.

Luckily, scientists have figured out other ways to determine Saturn's rotation rate that don't involve the magnetic field; we'll see how in the next chapter.

CHAPTER 5

How We Peer Inside Planets

ONE OF THE REASONS I came to Johns Hopkins from the University of Toronto back in 2016 was because of the greater opportunities to participate as a scientist on NASA missions. I got my first opportunity when the Mars *InSight* lander (the acronym for Interior Exploration using Seismic Investigations, Geodesy and Heat Transport) touched down on Mars's surface in November of 2018. The landing site was just north of Mars's equator in a flat region called Elysium Planitia. (*Elysium* is the underworld of Greek mythology, while *planitia* is Latin for "low plain.") *InSight*'s landing site is about 370 miles north of where the Mars *Curiosity* rover has been driving around in the Gale Crater since 2012 (still operating at the time of this writing), searching for signs that Mars once had a habitable environment.

InSight was the first Mars mission dedicated to studying the planet's interior; I was excited to learn more about the red planet's core, as its magnetic field is somewhat of a mystery. It doesn't have an active dynamo today generating a global-scale magnetic field like Earth does, but the rocks in most of Mars's crust are magnetized and indicate that it had a dynamo very early in its history. We don't know exactly when, but looking at which rocks are magnetized and their ages suggest the dynamo operated sometime between 4.5 and 4.1 billion years ago *and* between 3.9 and 3.7 billion years ago.[1] Notice the gap of 200 million years between those intervals: We think the dynamo may have been "off" during that time because some of the largest impact craters on Mars that formed between 4.1 and 3.9 billion years ago, like the Hellas and Argyre basins, don't show signs of magnetization in their rocks.

Why would a dynamo shut on and off? Does that happen on other planets? Could it happen on Earth? If so, what would the implications be for life on Earth? The first step to answering these questions is to learn as much as possible about the structure, composition, and evolution of Mars's interior. That will help us understand when its core could have been convecting in order to generate a dynamo.

This is where the InSight mission comes in. The mission's payload—the various items of equipment onboard—comprises a range of instruments to collect data on Mars's interior, including a seismometer and a radio antenna. The seismometer was designed to detect quakes. The radio antenna was built to track Mars's rotation (as well as communicate with Earth). What would quakes on Mars and rotation tell us about Mars's interior? Stay tuned as we explore these areas of inquiry.

SEISMOLOGY ON EARTH

Living on Earth definitely has its advantages. There's the breathable atmosphere, abundance of food, and magnetic shield, but for someone who studies a planet's insides, another advantage is the ability to measure earthquakes.

To understand more about Mars from its quakes, it helps to look at what we've learned about Earth from detecting earthquakes. An earthquake occurs when part of the Earth's lithosphere suddenly shifts relative to another part. These shifts happen because the lithospheric plates move relative to each other on the surface and because the plates can deform by being pushed and pulled on by other plates, creating features like mountains and rift valleys. In the United States, this happens frequently in California, the site of lateral motion between the Pacific and North American plates, for example, along the San Andreas fault. The sudden shifting of rock causes waves to travel from the earthquake source through Earth.

With a seismometer, scientists accurately measure how big the waves are (their amplitudes) and the time of their arrival at the seismometer location. The amplitudes tell us about the strength of the earthquake, but the timings of the waves' arrivals are particularly important for understanding a planet's interior.

The closer you are to the source of the earthquake, the sooner the seismic waves reach you. Using arrival times from seismometers distributed over the Earth, therefore, helps to determine the location of the earthquake. Importantly, these waves have traveled *through* the Earth, and their speeds are directly related to the material properties through which they traveled. That means we can use the timings of the waves' arrivals to determine material properties like density, phase, and viscosity. From these measurements, we can infer what the compositions, pressures, and temperatures are deep in the Earth.

Depending on the seismometer's location and the earthquake source location, different parts of Earth's interior will have been traversed by the waves. We can therefore determine detailed local information on what makes up Earth's interior by looking at global maps of wave arrival times from earthquakes.

This technique allowed scientists to determine the radius of Earth's core in 1906. That's because seismic waves travel faster in denser materials, like the iron in Earth's core. It's also how we know that the outer two-thirds of Earth's core is liquid, because certain types of seismic waves (called S waves) can't travel in liquids and end up reflecting or transforming at the interface of a solid and liquid.

With sophisticated data analysis techniques, seismologists can now determine tiny variations in material properties inside the Earth. One of numerous research projects focused on this is the Global Seismographic Network, which consists of 152 permanent state-of-the-art seismographs distributed across the

Earth's Seismic Waves

When an earthquake occurs, it radiates seismic waves from its location (known as the epicenter). Two types of waves are generated that travel through Earth. P-waves ("primary" or "pressure") are akin to sound waves traveling through Earth. The waves compress and dilate the material as they move through it. S-waves ("secondary" or "shear") are akin to waves that travel along a rope or string when you pluck one end. The wave shears the material laterally from the direction the wave is traveling.

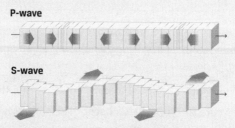

P-wave

S-wave

Image source: Dr. Edward Garnero, Arizona State University

S-waves and P-waves travel through Earth and arrive at surface locations at different times based on the distance from the epicenter. The speeds of the waves are determined by the physical properties of the material they travel through.

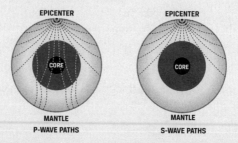

P-WAVE PATHS

S-WAVE PATHS

globe that share real-time data with the research community.[2] In addition to seeing the fluid and solid core interfaces, seismologists have found other boundaries in the mantle where minerals change phase. They've also detected an enigmatic layer at the bottom of the mantle, called the D″ (pronounced "D-double-prime") layer, which appears to be anywhere from tens to hundreds of miles thick over the globe. Scientists have suggested that it's the location of the "slab graveyard"—that is, the place where the lithospheric slabs that descend into the Earth at subduction zones end up when they've reached the core-mantle boundary. Here they stay for a while until they heat up enough to become buoyant and rise again (just like soup heated in a pot) as part of the mantle convection process. But they've likely mixed a lot and experienced chemical reactions with the surrounding mantle during this time, and so what emerges at mid-ocean ridges as new lithosphere is not recognizable as its former self. The entire cycle takes hundreds of millions of years.

SEISMOLOGY ON OTHER PLANETS AND MOONS

If you asked most planetary scientists (or at least those who study rocky interiors) what instruments they'd most like sent to other planets, I'd bet they'd say seismometers. The richness of the information provided by seismic waves is hard to get from other measurement techniques such as those involving gravity and magnetism. But seismometers are challenging instruments to use on other planets. First, ideally, you want lots of them in order to get good spatial coverage across massive surfaces. Seismometers also need to be placed directly on the planet's or moon's surface to feel the waves. So far, we've only placed seismometers on four worlds, including Mars and the Moon. We've also attempted readings on Venus, which melts probes before they have a chance to perform much science. The upcoming Dragonfly mission, currently slated to reach Saturn's moon Titan in 2034, will also have a seismometer. Mercury, the last terrestrial planet we haven't sent a lander to, is difficult to reach because of the Sun's gravity, while the four giant planets don't have solid surfaces. We could send seismometers to the moons of the giant planets in the future, but it might prove an expensive undertaking, even by NASA standards. Contrast that to magnetic and gravity measurements, which can be made from a single instrument in orbit around a planet. You also need to shield the seismometers from shaking due to wind if you want to isolate the waves coming from the interior from those coming from the atmosphere. And then of course you need earthquakes–or, I guess, "marsquakes" or "moonquakes" or "venusquakes" or . . . you get the picture.

On Earth, these quakes mainly come from the movement of Earth's lithospheric plates as part of plate tectonics. Although plate tectonics don't happen today on any of the other planets,[3] quakes can be produced through a few other processes. First, even though other planets don't have plate tectonics, their surfaces and interiors can shift and deform through the creation of volcanoes or the cooling and shrinking of the planet. Tidal forces can also stretch and bend a planet, causing quakes.

SEISMOLOGY ON THE MOON

The Apollo missions which first visited the moon in 1968 are legendary. They began humanity's voyage beyond Earth—I like to think that standing on the barren surface of the Moon in a giant white spacesuit, breathing oxygen through a tank, and looking back at Earth through a visor provides a strong reminder of how precious and unique the environment of our own planet is for humans. The experience caused Apollo 14 astronaut Alan Shepard—the first American to reach space—to remark in his book *Moonshot*, "But from here, from the moon, [Earth] is, in fact, very finite, very fragile . . . so incredibly fragile. That thin, thin atmosphere, the thinnest shell of air hugging the world—it can be blown away so easily! A meteor, a cataclysmic volcano, man's own uncaring outpourings of poison . . ."[4]

Aside from planting flags, collecting rocks, roving around in a lunar buggy, and playing some epic golf, the Apollo astronauts also carried out a range of scientific experiments. This included installing seismometers at each of the Apollo landing locations.

Neil Armstrong and Buzz Aldrin set up the first seismometer in the Mare Tranquillitatis—the Sea of Tranquility (thus named by early astronomers who mistook the basaltic plain for a sea), not far from the Eagle lander. Amazingly, the seismometer almost immediately began recording moonquakes, proving that Earth wasn't the only planet with quakes and that the Moon is a dynamic entity.

The lunar seismometers recorded more than 12,000 moonquakes between 1969 and 1977. Some of these were caused by the shifting and cracking of the Moon's crust because of thermal stresses as regions of the Moon heat up and cool down—a temperature swing of 400 degrees Fahrenheit—as they come in and out of view of the Sun.[5] Other moonquakes were caused by meteor impacts onto the surface. Interestingly, some moonquakes had sources that were very deep in the Moon and repeated every 27 days. That's the length of the lunar year (i.e., the time it takes the Moon to orbit the Earth). Researchers determined that these quakes were caused by tidal forces stretching and squeezing the deep interior of the Moon.

In 2011, 34 years after the *Apollo* seismometers were turned off, a group of scientists reanalyzed the seismic data from the Apollo missions using modern computational techniques, as the science has naturally evolved. Using these new data processing approaches, they found evidence for the location of the core-mantle boundary in the Moon, confirming that the Moon has a small 200-mile radius iron core.[6]

Although the *Apollo* seismometers were turned off in 1977, there's a future for seismology on the Moon. In 2021, NASA selected the Farside Seismic Suite to be an instrument package delivered to the Schrödinger Crater on the far side of the Moon around 2025 as part of the NASA PRISM (Payloads and Research Investigations of the Surface of the Moon) program.[7] This experiment will provide a better understanding of the deep interior of the Moon, revealing whether there are any differences in seismic events between the nearside and farside of the Moon.

The Moon was humanity's first foray into planetary seismology beyond Earth and, happily, not our last. The next stop for planetary seismology was Mars.

MARTIAN SEISMOLOGY

Aspirations for recording and examining seismology on Mars are about as old as seismology on the Moon. In July 1976, the Mars Viking 1 mission, the first spacecraft to land on Mars, carried a seismometer. The plan was for the seismometer to sit on the deck of the lander and sense marsquakes through their effects on the legs of the lander. Unfortunately, the system in place to unlock the seismometer for use didn't operate correctly, and the seismometer couldn't function. There aren't exactly mechanics we can send to other planets to fix malfunctioning instruments, at least for right now.

Luckily, the *Viking 2* lander wasn't far behind *Viking 1*, landing on Mars at a different location two months later. This time the seismometer unlocked (yay!), but planetary scientists learned that trying to record marsquakes traveling through the lander legs was not feasible (boo!). That's because the seismometer recorded shaking from everything, not just marsquakes. So anytime the spacecraft antenna had to be rotated or the robotic arm or cameras were moved—basically anything the spacecraft did—the seismometer recorded shaking signals. Winds up to 70 mph were also a problem, shaking the spacecraft and causing more signals in the seismometer. So back to the drawing board, so to speak. Scientists learned that if a seismometer were to be deployed on Mars, it had to be off the spacecraft, in touch with the ground, and shielded from wind. This was an important lesson, and it inspired the design of the ambitious Mars InSight mission.

InSight landed on the Martian surface in November 2018. After some housekeeping tasks, like unfurling the solar panels and taking pictures with the cameras to determine the surroundings, it was time to get work on the seismometer. It needed to be moved from the top of the lander to the ground so that it could directly detect marsquakes. That was accomplished with a robotic arm gripping the seismometer and carefully placing it on the ground. Then a windshield was moved from the top of the lander to cover the seismometer, making it harder for those pesky winds to create false signals.

In April 2019, *InSight* detected the first marsquake. I remember this time as a participating scientist on the mission, and I think all team members simultaneously cheered and also let out a huge sigh of relief. Getting that seismometer functioning on Mars was an epic task, and massive kudos go to the global team of engineers and scientists who made it happen. By the time the mission ended in January of 2023, a total of 1,319 marsquakes had been detected. Many of the marsquakes are caused by impacts, but some are also caused by faulting along surface fissures in the Cerberus Fossae graben system (a graben is a block of crust surrounded by faults where the crust is being pulled apart) about 1,000 miles away from *InSight*. There are signs of flowing liquids (either magma or water) in that region in the very near past—as recently as 50,000–200,000 years ago—which for a geologist is *very* recent.

The first sets of marsquakes were a bit disappointing for someone like me interested in the deep interior of the planet (I was, however, very happy for those more interested in the shallower layers of Mars, like its crust and upper mantle). The quakes weren't very big (typically smaller than magnitude 4 on the Richter Scale) and didn't have the necessary parameters to enable the detection at the lander of seismic waves that had traveled through Mars's core. But on the 173rd Martian day of the mission (which is about 178 Earth days), a quake with the right location and magnitude (around magnitude 5) was detected. Several others were detected as the mission continued, and the data from these larger quakes revealed the deep interior of Mars.

From these marsquakes, scientists pinned down the radius of Mars's iron core (1,137 miles), which was on the upper end of previous estimates. This larger core size also means that the core is somewhat lighter than previously thought, with a larger fraction of elements such as sulfur, oxygen, and hydrogen mixed in with the iron. Figuring out the exact composition is a work in progress and will require new laboratory and computer studies of mixtures of iron with lighter materials to determine their geochemical behavior. Determining the exact composition will be very insightful (see what I did there) for understanding the building blocks of all the terrestrial planets, including Earth. This will ultimately lead to understanding when Mars's core stopped convecting and hence when its dynamo died.

GIANT PLANET SEISMOLOGY

At first it might seem odd to think of doing seismic measurements on a giant planet. After all, one would think that a major requirement for seismology would be the presence of a solid surface for the seismometer to sit on while recording the ground motions. But seismic waves can also occur inside giant gas and ice planets as well as the Sun, caused, for example, by impacts or convective motions in the planet's interior. These waves can emanate through the planet, sort of like how a bell rings, and the oscillations can reach the atmospheric layers, causing them to bob up and down. We can't currently fasten a seismometer to the gaseous outer layers of a giant planet, but we can visually observe the bobbing of the surface and use that to infer details about the waves.

This procedure was first done for the Sun. Called helioseismology, the study of the waves ringing in the Sun has provided information on the Sun's internal structure and motions. For example, we now know what the internal rotational profile looks like in the Sun: The inner 70% (roughly) of the Sun's radius rotates like a rigid body (i.e., all at the same rotation rate), but the outer 30% of the Sun's radius experiences different rotation rates at different latitudes and depths. This "differential rotation" combines with convective motions in this layer and is ultimately responsible for the generation of the Sun's magnetic field and sunspots.

To date, we haven't been able to do the equivalent of helioseismology for giant planets, although theoretically it's possible, and scientists are working on this. The surface ringing of planets is much more muted than that of the Sun, making it a harder measurement to make. However, one of the giant planets, Saturn, has provided a novel sort of seismology used to study the planet's interior, and it's all thanks to its splendid rings!

SATURN RING SEISMOLOGY

In my humble (yet correct) opinion, Saturnian rings are the most beautiful sight in the solar system. You can see them from

Earth with a telescope, and indeed this is how Galileo first observed the rings in 1610. But to truly appreciate the rings, you need to get up close, and this is exactly what the Cassini mission to Saturn did from 2004 to 2017.

The rings span about 45,000 miles from their inner to outer edge and are extremely thin (only 30 to 3,000 feet thick). To put that in perspective, if you were to look at them edge on, their thickness to length ratio is less than the same ratio for a sheet of paper. The rings are mainly composed of water ice particles ranging in size from specks to boulders tens of feet across. Every single particle in Saturn's rings is orbiting the planet, and that means they can act as sensitive detectors of density variations inside the planet (just like the GRACE and GRAIL orbiters detected variations in Earth's and Moon's gravity fields, respectively).

In the novel 2312 by Kim Stanley Robinson, some of the characters go surfing on the waves in Saturn's rings (lucky them). Although the technology to do so may be hundreds of years in the future, the waves in the rings are real, and we can see them today. Some are caused by gravitational interactions with tiny moonlets in the rings, but others are caused by waves resonating inside Saturn itself. Deep in Saturn's interior, waves resembling sound waves travel back and forth. This causes density perturbations as the material in the waves bob up and down. The particles in Saturn's rings notice these density perturbations by the small changes in gravity they cause, and this causes the ring particles to oscillate, propagating waves through the rings themselves. Looking at the pattern of the waves allows scientists to infer properties of Saturn's deep interior.

Saturn ring seismology has told us that the inner 40% (roughly) of Saturn is a stably stratified fluid. Although we think of Earth as having separate "mantle" and "core" layers, the deepest part of Saturn can't be separated into layers. Instead, the fraction of heavier compounds, like ices and rocks, in Saturn must keep increasing with depth in Saturn's core. That means there's more rock and ice closest to the center and continuously less as you move outward, where it gets replaced with more hydrogen and helium. It's this gradient in composition that makes the deep interior stably stratified. We now call this scenario a "fuzzy core."[8]

And Saturn's not the only planet believed to have a fuzzy core. Although not determined from ring seismology, the Juno mission to Jupiter discovered that its core must also be fuzzy and hence not made up of separated layers. It appears that fuzzy cores may be a natural state for gas giant planets and a result of how these planets formed and evolved over the past 4.6 billion years.

BACK TO SATURN'S ROTATION PERIOD

In the previous chapter, we discussed how we can't use Saturn's magnetic field to determine the planet's rotation period.

Remarkably, scientists figured out how to use these ring waves to do so instead. Using ring waves, which are following waves in Saturn's deep interior, they find Saturn's rotation period to be 10 hours, 33 minutes, 38 seconds (so Saturn's day is less than half as long as Earth's day).[9] That number is about six minutes shorter than the value (mis)determined from Saturn radio emissions by the Voyager missions and 12 minutes shorter than the (mis)determined value from radio emissions observed by the Cassini mission. It appears Saturn is rotating a bit faster than we initially thought!

STUDYING A PLANET WITH ITS ROTATION

Now that we know how fast Saturn is rotating, we can actually use that rotation to study properties of its interior. This works for other planets, too. The fact that a planet rotates can have major implications for its shape and dynamics, and is also responsible for the swirly-ness of hurricanes, ocean gyres, and tornadoes.

First, planets develop a bulge at the equator because of their rotation. The parts of the planet furthest from the rotation axis (i.e., at the equator) have faster speeds as they spin around the rotation axis than the parts closer to the rotation axis (such as high latitudes). The centrifugal forces on the equatorial parts are therefore bigger and cause the equator to shift outward a bit, creating that bulge.

A planet's rotational bulge tells us about the mass distribution inside the planet. If a planet were a perfectly uniform sphere—that is, the density was the same everywhere—then we could predict how bulgy the planet would be given the competing effects of self-gravity and the centrifugal force caused by the rotation. But when we do that for the giant planets, it turns out that they're all bulgier (larger in the equatorial direction) than what would be predicted for a uniform sphere.

Take Saturn, for example. It's the bulgiest of the planets, with the equatorial axis about 10% wider than its polar axis. This result tells us that Saturn has denser material concentrated toward the center of the planet and lighter material in its outer layers.

We can also use the size of the bulge to predict Saturn's rotation period (and indeed this was done before the ring wave determination for the rotation period). The amplitude of the bulge is determined by how fast the planets are spinning— the faster they spin, the bulgier they are. Using information about the density profile in Saturn determined by gravity measurements, scientists could use the size of Saturn's equatorial bulge to determine the rotation period. With this technique, Saturn's length of day was found to be 10 hours, 32 minutes, 35 seconds,[10] which turns out to be very close to the value found from ring waves. This gives some confidence that the rotation periods we're measuring from ring waves and from the rotation are correct.

The rocky planets are much less bulgy than the giant planets, but they still have an equatorial bulge. The distance from the center of the Earth to the equator is 3,959 miles, whereas the distance from the center to the pole is 3,950 miles. That's about a 0.2% bulge compared with Saturn's 10%.

But some interesting results can be found when looking at the rotational bulge of rocky planets. My favorite example comes from looking at our Moon, which, while a satellite, has the most in common with the terrestrial planets of the inner solar system. It turns out that the Moon is too bulgy. That is, the Moon's equatorial bulge is much bigger than it should be given its current rotation period. This mystery is solved by looking at the Moon's history.

As discussed in chapter 2, the Moon formed from a giant impact that put a disk of material in orbit around the Earth. A few decades afterwards, that material left in the disk had coalesced into the Moon. It would have been fascinating to watch that process happen, especially since the Moon would have looked very different in the sky at that time. First, it wouldn't have any of those big impact craters we see as dark circles on the Moon (that's because those formed around 400 million years after the Moon formed). But perhaps more importantly, the Moon would have been *huge* in the sky. That's because the Moon formed much closer to Earth, and was also spinning much faster. Over time, the tidal interactions between the Moon and Earth have worked to slow down the Moon's rotation speed and move it farther from Earth. In fact, the Moon is *still* receding from Earth today at a rate of about one inch per year.

So how does the Moon's formation and evolution explain the Moon's extra bulginess? When the Moon initially formed, it would have been in a molten state, eventually cooling to form a solid surface. The shape of the Moon was determined by the bulge that was frozen in when that happened. And the Moon was spinning faster back then. The Moon's current shape therefore contains a "fossil bulge" from billions of years ago.

PRECESSION AND NUTATION

Rotation not only affects the shape of a planetary body (i.e., its equatorial bulge); it also affects how the rotation axis of the body moves in space. If you've ever played with a spinning top, you've witnessed some of these motions. The top, which is usually top-heavy, and then ends on a point at the bottom, is set spinning by a flick of your fingers and let loose on a surface. This spinning gives the top angular momentum, and it would keep spinning at the speed you gave it if it weren't for pesky friction and air drag.

If you give the rotation axis a slight tilt when you spin the top (i.e., don't keep the top perfectly vertical), you'll notice a curious behavior. The axis of the spinning top, which is tilted over, doesn't stay fixed, but it doesn't fall over, either (which is what it would do if it weren't spinning). Instead, the axis traces a circle out in space. This motion is called "precession" and it happens as a result of Earth's gravity torquing the top while it's spinning.

A planet can also be torqued by gravitational forces from other nearby planets, moons, and even the Sun because of equatorial bulges. Take Earth, for example. The equatorial bulge of the Earth, along with the tilt of the axis with respect to the orbits of the Moon and Sun, means that the Moon and Sun torque the Earth slightly. This causes Earth to *precess:* the orientation of the Earth's rotation axis in space changes in time. It takes about 26,000 years for the precession to complete one cycle.

Today, the north star is Polaris because that's the star right above Earth's geographic (or rotation) pole. But the north star won't always be Polaris. As Earth precesses, the rotation pole will move in space, and different stars will end up located over the north pole. In 12,000 years, almost half-way through the precession cycle, the north star will be Vega, a very bright star in the constellation Lyra (and also where the alien signal was coming from in the movie *Contact*, based on the book by Carl Sagan).

There's another motion you can get from a spinning top if you try the following: While the top is spinning, flick its highest point with your finger as if you were trying to knock it over. This will cause the rotation axis to wobble back and forth a bit as it precesses. This motion is called nutation.

The precession and nutation of a toy top is somewhat simple because the top is simple. But planets are sometimes complicated. For example, they might have liquid layers. Spinning something with a liquid layer inside looks quite different than with a solid object. If this seems strange, try the following experiment: Take a hard-boiled egg (still in its shell) and a raw egg (also still in its shell) and place them on your kitchen counter. Using your hand, give them both a spin. You'll notice that the raw egg, with its liquid layers, spins very differently than the solid hard-boiled egg. The same is true for planets.

Earth's rotation axis precesses and nutates slightly differently than it would if it were completely solid. If we didn't have seismology to tell us Earth has a liquid layer in the core, tracing out the rotation axis in space would.

The InSight mission at Mars also used this technique to study Mars's rotational motion with the RISE experiment (Rotation and Interior Structure Experiment). By using the radio antenna on the spacecraft put there for us on Earth to communicate with the lander, scientists were able to closely track the motion of the lander on the surface of Mars. Now, the lander wasn't moving *on* the surface, the surface itself was moving, with the precessional and nutational motions. The RISE experiment actually discovered that Mars's iron core is liquid, independently and before the seismic experiment did.[11]

Before leaving this discussion of the InSight mission, I feel the need to mention the InSight HP^3 experiment (acronym for Heat Flow and Physical Properties Probe). Recall from chapter 1 the fervent effort to drill into Earth's interior to discover the properties of the crust and mantle. InSight planned to do the same on Mars with a drill affectionately called the mole. The drill was designed to burrow about 10 feet down into Mars. From that point it would take temperature measurements to determine how much heat was coming out of Mars. This would help to answer questions about how fast Mars was cooling, and how it had evolved in time.

Chapter 1 demonstrated how hard it can be to drill into Earth. It becomes exponentially harder when you're trying to drill into another planet for a multitude of reasons: no humans around to fix stuff, little knowledge of the properties of the crust, and so on. Unfortunately, the mole didn't manage to reach 10 feet deep because the soil layer of Mars didn't end up having the friction necessary for the drill to work. After a herculean effort by the scientists and engineers on the InSight team, the mission had to accept that the mole would only make it an inch or so down. However, it was still able to make some measurements of the thermal properties of the soil in the region. Lessons were learned that can be applied to future missions, and we were reminded once again that space missions are hard.

MOMENT OF INERTIA

Studying a planet's precession is also helpful because the precession period is related to a fundamental parameter of a planet: its moment of inertia. This is a weird parameter to explain and is (perhaps) easiest to understand when compared with another fundamental parameter of a planet: mass. Mass tells you how much force you need to accelerate an object. That's from Newton's second law (force = mass × acceleration). Basically, the more massive an object, the more force you need to accelerate it. The moment of inertia tells you how much torque you need to accelerate the rotation of an object. The larger the moment of inertia, the more torque—the force that causes an object to rotate on an axis—you need to change the rotation rate of an object.

Calculating an object's moment of inertia can be complicated because it depends on the mass of the object *and* on how that mass is distributed around a rotation axis. For example, a solid sphere with a mass of 10 kilograms has a different moment of inertia than a hollow sphere with the same radius and same mass. The hollow sphere has a larger moment of inertia because more of the mass is located farther from the rotation axis. If the spheres were rotating, you would have a harder time changing the rotation of the hollow sphere than you would the solid sphere.

We can learn about how mass is distributed in a planet by measuring its moment of inertia. If most of the mass is concentrated near the center, say in an iron core, then the moment of inertia is smaller than if the density is uniform throughout the planet. Because planets are very spherical, we typically compare their moments of inertia with what they would be if they were spheres of uniform density. All planets increase in density with depth because of the effects of pressure, so no planet has a larger moment of inertia than a uniform sphere.

Luckily, the moment of inertia of a planet is directly related to the rate of precession of the rotation axis, which we can measure. The moment of inertia has been determined for the planets and many of the larger moons in the solar system. Using their moments of inertia, we can learn a lot about how different two planetary bodies are even if they look quite similar from the outside.

As an example, consider Jupiter's moons Ganymede and Callisto. They're very similar in size, density, and composition but are very different in their moments of inertia. Ganymede's moment of inertia is about 77% of what a uniform sphere would be with the same mass and radius as Ganymede. Callisto's moment of inertia is about 89% of a uniform sphere of the same mass and radius as Callisto. Ganymede's lower percentage means that more of its mass is concentrated at its center. That tells us that Ganymede has differentiated, with an iron core at the center, surrounded by a rocky mantle, and then a thick ice layer surrounding the rock. Although Callisto is very similar in terms of bulk composition, its larger moment of inertia percentage compared with a uniform sphere means that it hasn't differentiated as much as Ganymede. It still has an icy outer layer, but deeper down, the interior is a mixture of ice, rock, and iron that hasn't fully separated out.

We don't completely understand why these moons have such different interior structures. They formed around the same time from the same disk of material that surrounded Jupiter. The only difference is that Callisto is slightly farther away from Jupiter than Ganymede is. For some reason, that means Callisto's interior didn't heat up and melt as much as Ganymede did while they formed. It's this melting that allows the bodies to differentiate.

One interesting theory for why Ganymede was heated much more than Callisto during their formation has to do with meteor impacts these moons experienced around 4 billion years ago.[12] Early in the solar system, more impacts were occurring because there were more planetesimals and asteroids around. Because of Jupiter's large mass, asteroids that get near it can get trapped in its gravitational well and accelerate toward the planet.

Ganymede is closer to Jupiter than Callisto, and this has two effects. First, Ganymede is more likely to get hit by meteors than Callisto. That's because Ganymede has a smaller orbital radius and a faster orbital speed—akin to your increased odds to get hit by a car driving faster and on a smaller circular track than one on a slower and larger track if you were foolishly trying to cross the tracks.

Scientists predict that Ganymede was hit with twice as much material from impacts as Callisto. Second, the meteors hitting Ganymede would be moving faster (and hence have more energy) than the meteors hitting Callisto. This higher energy of impacts and more impacts mean that it's likely that Ganymede was heated by impacts much more than Callisto was. This could explain why Ganymede fully differentiated while Callisto only partially differentiated.

A RANT ABOUT VENUS

To end this chapter, I'm going to briefly rant about Venus. For a scientist studying planetary interiors, Venus is the most frustrating planet. It does everything possible to stop us from using the techniques outlined in this and the last chapter to study its interior. Here are some examples.

First, Venus has no moon. That means we can't use Kepler's third law to quickly determine Venus's mass from the orbital period and distance of a moon. Then there's that thick, opaque atmosphere that makes it hard to see down to the surface and isolate a surface feature to watch spin around the planet in order to determine its rotation period. And because of the slow spin, Venus doesn't have much of a rotational bulge. Combine that with the incredibly tiny tilt of its rotation axis, and its precession rate becomes very hard to measure. That means it's hard to determine Venus's moment of inertia!

That atmosphere also makes it impossible to orbit Venus very close to the surface. This limits the detection of detailed gravity anomalies to unveil detailed density variations. And then, the planet has no active dynamo. Admittedly this does tell us something about its core—it's not convecting vigorously enough for a dynamo—but it's hard to determine why. Add to this the fact that the nearly 900 degrees Fahrenheit surface temperatures mean that it's really hard for the rocks in the crust to record the magnetic fields from any possible past dynamo action when they formed. So we can't even learn if Venus had a dynamo in the past, like the Moon and Mars did.

Finally, and perhaps most annoyingly, Venus' surface—with its 92 bars of pressure, scorching temperatures, and corrosive atmosphere composed mainly of carbon dioxide under high pressure—makes it currently impossible for machinery like seismometers to survive long on the surface, so studying Venus through venusquakes is also (to date) impossible.

My rant is now over.

Throughout the last few chapters we've explored methods we can use to learn about planetary interiors. Now it's time to focus more on what we've learned from those methods, including some science fiction–worthy characteristics of planetary interiors.

CHAPTER 6

Curious Planetary Elements

WE'VE ALL BEEN ASKED AT SOME POINT what our favorite food is. My answer varies, and it depends strongly on how well the said food is prepared. But I don't think I've ever been asked: "What's your favorite ingredient?" That one might be a bit harder to answer since the beauty of food, to me, comes in the blending of different ingredients and subsequent preparation and cooking to achieve perfectly balanced flavors and textures in a dish.

The same might be said for planets. Many of the fascinating features of planets come from the blending of different compositions that are then placed in a unique pressure and temperature environment (i.e., "cooked") for a given amount of time. For example, a chunk of iron at Earth's surface can't create a dynamo, but the same chunk of iron at the temperatures and pressures of Earth's core can.

One of my favorite food experiences is to encounter an ingredient used in an unexpected way in a dish. Admittedly, this can also be one of my least favorite food experiences depending

CURIOUS PLANETARY ELEMENTS   **149**

on how successful it is. Consider chocolate: Red chilis in chocolate? Not too surprising, but I'd like ten pounds of that right now, please. White chocolate in a pâté? Unexpected, yet delightful. But bacon in chocolate? What on earth were they thinking?

In this spirit, this chapter highlights some planetary ingredients and their unexpected yet delightful properties when cooked in the environments of their planets' interiors.

HELIUM AND PLATE TECTONICS

It's spring of 2022 and I'm about to send the weirdest email I've ever written. I'm chair of Johns Hopkins University's Department of Earth and Planetary Sciences, and several of the department's faculty members, who do really cool research on Earth's geological and geochemical history, have alerted me to an issue they're facing—they're running out of helium. Liquid helium, to be precise.

Liquid helium is used in a variety of vital laboratory instruments, like mass spectrometers and magnetic resonance imaging (MRI) machines, because these instruments employ superconducting magnets that need to be cooled to extremely low temperatures to work. Giant dewars of liquid helium are delivered to scientists' laboratories regularly, coming in capacities of anywhere from 30 liters to 1,000 liters. But that spring, a series of compounding events made some research laboratories come to a grinding halt. The superstorm of supply chain issues from the COVID-19 pandemic and unplanned shutdowns of US and international liquid helium plants were the causes of the scarcity.[1] And if you *could* get it, it was up to ten times more expensive than usual.

So there I was writing an email to a Hopkins colleague in another department who told me they had some liquid helium. "Who's your supplier? Any chance you can spare a bit? I can pay!"—and "I'll make it up to you later if we can get some and you need it then," were phrases I expected to hear around back alley deals, not in my regular workings as a science department chair.

Where does this helium that we rely on so much come from? It all goes back to planetary formation. The protoplanetary disk from which the planets formed had significant amounts of helium. It's the second most abundant element in the universe after hydrogen, accounting for about 25% of all regular (i.e., not dark) matter, while hydrogen is 70%. But the Earth never reached the critical mass needed to attract the gas in the protoplanetary disk to create a hydrogen- and helium-dominated atmosphere. That was only possible for the giant planets in the outer solar system.

Most of the helium we have here on Earth is actually created *inside* the Earth when radioactive elements like uranium and thorium naturally decay into other elements and release helium nuclei called alpha particles in the process. Although these radioactive elements only make up a few parts per million of Earth's composition, their radioactive decay is responsible for about half of the heat emanating from Earth.

This heat is critical for driving mantle convection in the Earth—which means it's also important for plate tectonics. Because of heat from radioactive decay, and leftover heat from Earth's formation stored inside the planet, the rocks near the bottom of the mantle are hotter than those near the top. Those rocks thermally expand making them lighter than their surroundings and so they buoyantly rise through the mantle.

At the same time, rocks that solidify from magma at the surface of Earth cool and become denser than their surroundings. This gives them negative buoyancy, causing them to sink through the mantle. Those rocks eventually make it to the bottom of the mantle, where they warm up again and become the hot rock rising through the mantle. This is the cycle of mantle convection—nature's recycling.

When I teach mantle convection, I tell students there's one concept I absolutely need them to understand and carry forth into the world when they meet new people: The mantle is *solid*. All of the convection processes, with rocks rising through the Earth and descending back down, happens with solid rock. I think there's a major misconception out there that the mantle is liquid, and it probably arises from the fact that when mantle rocks reach the surface of the Earth, like at volcanoes or mid-ocean ridges, the rock is in the form of molten lava (or magma if it forms under the surface). It turns out that rock only becomes molten when it gets close to Earth's surface, and it's because the rocks experience depressurization. Basically, the high pressures in Earth's interior keep rocks solid, even if their temperatures are several thousands of degrees Fahrenheit (well above the melting temperature of rocks at Earth's surface). But when that rock gets near the surface, the pressure is low enough for the rock to melt, forming the glowing orange ooze we witness as volcanoes erupt.

It may be hard to grasp that a solid like the mantle can convect. And this gets to a fundamental issue we face in studying Earth's interior: we don't have experiences with the properties of materials at high pressures and temperatures. Rocks can "flow" when they're hot enough, even if they're not melted. Glaciers, where solid water ice flows over mountains, are perhaps a good analogy. Admittedly, the flow of rocks in the mantle is extremely slow. A rock rising from the bottom of the mantle as part of mantle convection takes about 200 million years to reach the surface. That's equivalent to a speed of about half an inch per year (mantle rocks are definitely not winning any speed races).

Mantle convection is a necessary condition for plate tectonics, but it's not sufficient. All the rocky planets likely experience mantle convection, but only Earth has plate tectonics. Plate tectonics implies that the lithosphere is directly involved in mantle convection. The tectonic plates plunge into Earth's interior at subduction zones, and hot rock ascends through the mantle, creating new ocean floor at mid-ocean ridges. These processes are part of the convection in mantle convection.

In this way, the helium we use in our scientific instruments, as well as our party balloons, is actually a tracer of one of the main processes responsible for making Earth habitable to humans. Without this heat source, Earth's mantle would be much colder and more viscous; hence convection and plate tectonics would be harder.

HIDDEN HELIUM

Helium has also been at the heart of a mystery regarding Earth's interior for a long time. We seem to have too much of a particular isotope of helium: ^{3}He. Recall from chapter 3 that an element, like helium, can have different numbers of neutrons in its nucleus, and this creates different isotopes of the element. Essentially all of the helium we find in the universe (99.9998%) is ^{4}He, which has two protons and two neutrons. The rest (0.0002%) is ^{3}He (with two protons and only one neutron). ^{4}He can be either "primordial" (i.e., created shortly after the Big Bang) or produced from radioactive decay of other elements (like what happens with uranium and thorium in the Earth). But the tiny amount of ^{3}He out there in the universe is all primordial.

We know Earth didn't accrete any appreciable amount of gas from the protoplanetary disk when it was forming, so we don't expect it to have any primordial helium. We do expect Earth to have ^4He (from the decay of those radioactive elements inside Earth), but we shouldn't have any ^3He. The problem is, we do. When measurements are taken of the lava spewing out of volcanoes and mid-ocean ridges on Earth, it contains some ^3He.[2]

Admittedly, Earth could have accreted a bit of the gas from the protoplanetary disk, but the problem is, because helium is so light, as soon as it reaches Earth's surface, it gets transported to the high atmosphere and is then blown off the planet by the solar wind. You've experienced helium's lightness if you've ever held a balloon and tragically let go of the string outdoors.

We know that soon after Earth formed, it experienced melting and vaporization of crust and mantle rocks during giant impacts. And Earth is continuously recycling its mantle through plate tectonics. So, material from the depths of the mantle gets brought to the surface, and the surface descends back into the Earth at subduction zones—the boundaries where one plate moves down into the Earth under another. These events would have brought all the ^3He out of the Earth into the atmosphere to eventually be lost to space over the past 4.6 billion years. We shouldn't have any primordial helium left if we ever had any in the first place. Yet we still find some in those lavas! We know where the ^4He comes from—radioactive decay—but that can't explain the ^3He.

The detection of ^{3}He escaping from the Earth's surface led geoscientists to believe that some hidden reservoir of ancient material exists inside the Earth that hasn't participated in the plate tectonics cycle that would have removed the ^{3}He long before today.[3] Instead, ^{3}He generally stays confined to this hidden reservoir, offering only wisps of itself to the rest of the mantle once in a while. When that escapes to the surface, we detect it.

When I was in graduate school, I witnessed epic scientific debates about where this hidden reservoir was located. And of course, with anything that's hidden, how do you find compelling evidence for it? One scientific idea was that the lowermost part of the mantle didn't participate in mantle convection, and so only the "upper mantle" (i.e., the outermost few hundred miles of the mantle) convected. This would have allowed the lower mantle to hide things from our observation at the surface. This idea was dubbed the "layer cake" model of the mantle.

But the layer cake model seemed to contradict seismic evidence that the lithospheric plates that descend back into the mantle at subduction zones (what we call "slabs") do, in fact, make it down to the bottom of the mantle. In addition, seismic evidence implies that mantle plumes, where hot rock ascends through the planet, resulting in hot spot volcanic island chains like the Hawaiian archipelago, originate from near the bottom of the mantle. These plumes and slabs are the result of convection in Earth's mantle, and so it appears that the lower mantle does participate in convection and can't be the hidden reservoir.

That led scientists to propose the "plum pudding" model of the mantle (layer cake, plum pudding: clearly I'm not the only geoscientist obsessed with food analogies). In the plum pudding model, the hidden reservoirs weren't at the bottom of the mantle but instead, were found in clumps dispersed through the mantle, like the plums in a pudding. Updated versions of the layer cake and plum pudding mantle are still discussed today.

Another possibility is that the hidden reservoir is the core rather than the mantle. That has also been contentious as it's challenging to explain how the helium would have gotten down to the core in the first place, but scientists are working on this possibility.[4] I haven't seen this model given a food name yet, so I propose the "peach pit" model.

HELIUM RAIN AND METALLIC HYDROGEN

Let's expand our discussion of helium to other planets. Since helium is most abundant in the gas giant planets, Jupiter and Saturn, that's where we'll focus. As these gas giants formed, they accreted significant amounts of both hydrogen and helium gas from the protoplanetary disk. Indeed, to understand what happens to helium inside the gas giants, we need to consider it in the context of the accompanying hydrogen.

Our experience with hydrogen and helium on Earth is typically in their gas phase, unless they've been purposefully compressed into a liquid state for cooling instruments like those discussed in the previous section. In contrast, the hydrogen and helium inside Jupiter and Saturn are under enough pressure in their environments to exist naturally in nongaseous forms.

In the outer layers of Jupiter and Saturn, the hydrogen and helium are in gaseous phases, making up the bulk of these planets' atmospheres. But as we move deeper into the planets, pressures and temperatures increase. Eventually, the pressure and temperature become high enough for the hydrogen and helium to behave more like a fluid with a mix of liquid and gaseous properties.

As we go even deeper, more exotic phases of hydrogen and helium emerge. An important one for these planets' magnetic fields is the fluid metal phase of hydrogen. Perhaps it's best to review what it means to be a "metal." Metals are typically defined in terms of their physical properties: They're good at conducting heat and electricity, and they're malleable and shiny. And I should clarify that I am *not* using the astronomers' definition of a metal here, which is "anything other than hydrogen and helium," but rather the "real" definition one would use in chemistry, mineralogy, and material science.

The key to understanding metals is to grasp *why* metals have these properties. It has to do with how free their electrons are. In metals, the outer shell of negatively charged electrons that orbit an atom's nucleus are only weakly attracted to the positive nucleus and can become more attracted to surrounding nuclei. They also actively try to avoid the other electrons in the area. This causes the electrons to wander freely between the atomic nuclei, creating metallic bonds. If you take a nonmetal at room pressure and temperature—say, hydrogen—and squeeze it to high pressures, the atoms get closer and closer together. Eventually, they are so close that the electrons end up in this freed state with metallic bonds. That explains the high conductivity of metals (because the electrons can move freely, transporting charges or heat) and the malleability (it's easy to reshape them because the metallic bonds are so movable).

At a depth of about 10% of Jupiter's radius, and 30%–40% of Saturn's radius, the pressures are high enough for hydrogen to become metallic and conduct electricity. It's in this fluid metallic hydrogen layer that Jupiter and Saturn have dynamos that generate their magnetic fields. The fluid metallic hydrogen is convecting, just like Earth's core is convecting, and so magnetic fields are generated in these planets. These magnetic fields protect Jupiter and Saturn the same way Earth's magnetic field does.

But what about the helium? Jupiter's and Saturn's atmospheres are about three-quarters hydrogen and one-quarter helium (with trace amounts of other molecules). They're mixed together in a homogeneous fluid, that is, you can't really separate the hydrogen from the helium because it's all combined.

When this happens, we say that the materials are *miscible*. For everyday examples, think of salt water, where the salt is dissolved in the water. It's also possible to have fluids that are immiscible, like oil and water. If you try to mix them, they just separate out, with the denser one sinking to the bottom of the container.

Hydrogen and helium have a complicated relationship inside the giant planets. They're mixed in the outer layers like the atmosphere, but when the pressure gets high enough that hydrogen becomes metallic, the helium becomes immiscible. At this depth, droplets of helium condense out of the hydrogen, like water droplets condense out of the atmosphere onto the side of your cold glass in warm weather. Helium is denser than hydrogen and so the helium sinks, creating helium rain.

Deeper in the planet, the pressure and temperature conditions allow hydrogen and helium to become mixed again, so this helium rain only occurs in a layer (or shell) inside the planets.

SATURN'S ODDITIES EXPLAINED BY HELIUM RAIN

Although this helium rain is occurring deep inside the giant planets, there's evidence of it at the planets' surfaces, especially on Saturn. It turns out that Saturn has three oddities that can be explained by the presence of helium rain.

First, when you measure the amount of heat coming out of the planet, there's too much of it to be explained by a world that is fully convective and has been around for the past 4.6 billion years. What does that mean? When a planet (or a pot of water on the stove) convects, it removes heat. So planets cool down over time, and we can predict how much heat should be coming out of them today. For Saturn, there must be some ongoing source of heat in the interior, in addition to the heat of formation and radioactive decay, to explain the temperature of the planet. Having a layer inside Saturn where helium is raining out of hydrogen and sinking helps with this source of heat. Energy, in the form of heat, is released whenever a material sinks through another. It's like a kind of friction between the two materials. For fluids, we usually refer to frictional heating as "viscous heating" because it's affected by how viscous the fluids are. So the helium rain can explain the increased heat inside Saturn.

Second, Saturn's atmosphere seems to be missing some helium. As the planet formed and accreted the hydrogen- and helium-rich gas from the protoplanetary disk, we expect that the relative amount of helium accreted to Saturn would be the same as that in the disk. The disk was about 25% helium, so the gas Saturn accreted should also be about 25% helium. How do we know how much helium was in the protoplanetary disk? We look at the Sun, which was also accreting the same gas. We can also look at Jupiter. Jupiter and the Sun both have similar helium abundances. And although above, I said Saturn's atmosphere is "about 25% helium," the precise amount is a bit lower (although admittedly very hard to measure). Jupiter's value is much closer to the Sun's measured value of about 25%. So Saturn seems to have less helium in its atmosphere, where we can detect it. Saturn's atmosphere also seems to be missing an appropriate amount of the noble gas neon, which can be explained by helium rain. (On Earth, neon lights up restaurant windows and beer logo signs in dive bars.)

Some of the helium in Saturn's atmosphere makes it down to the depths where hydrogen becomes metallic. Neon is a gas that tends to combine with helium when helium is in liquid form. So the fact that some neon seems to be missing from the atmosphere as well suggests that it gets transported to the depths of Saturn from the atmosphere as well during the helium rain process. When helium gets down there and condenses, it falls deeper in the planet and has no means to make it back to the outer atmospheric layers. So the atmosphere ends up depleted in helium because the helium is stuck deeper inside the planet. It's sort of like cookies and milk. If you notice some cookies missing from your kitchen because they've been transported to someone's stomach, you might infer that you're also missing some milk, since it goes so well with cookies.

The third oddity is that Saturn's magnetic field has some unusual properties. It's dipolar, like Earth's field, but it's too perfect in a way (I'm sure all the other planets are jealous). We initially talked about this in chapter 4 in the discussion of Saturn's length of day. Saturn's magnetic axis is perfectly aligned with its rotation axis, meaning that the magnetic and geographic poles coincide. The Earth's (and other planets') magnetic poles are close together, typically within about 10 degrees in latitude (with the exception of Uranus and Neptune, which we'll discuss in the next section), but they aren't perfectly aligned the way Saturn's is. When we run computer simulations of how planets make magnetic fields, it's almost impossible to reproduce this feature of Saturn's field unless we include the physics of helium rain in the computer simulations. The helium rain layer in Saturn acts as a sort of filter that attenuates any non-axisymmetric features from the magnetic field so we don't see them from spacecraft. The magnetic field inside Saturn has these non-axisymmetries, but the fields we observe outside the planet do not.

A great question I'm asked a lot about helium rain is: "Why is this happening in Saturn and not Jupiter?" It turns out we aren't completely sure whether it's happening in Jupiter. Because Saturn is a bit less massive than Jupiter, it has cooled a little faster over time. If we look at the required pressure and temperature ranges that are needed for helium rain to form, they occur in a much thicker layer in Saturn than in Jupiter. It's possible that Jupiter has a helium rain layer, but it's much thinner than Saturn's helium rain layer and so hasn't had as much of an effect on observables (like the magnetic field, excess heating, and helium depletion). Some intriguing hints from the NASA Juno mission that's orbiting Jupiter are starting to suggest there really is helium rain in Jupiter as well.[5]

Core Detectives

In 2011, NASA launched the *Juno* spacecraft toward Jupiter; it arrived in Jupiter's orbit in 2016 and has investigated the planet ever since. Jupiter is the largest planet in our solar system, at 11 times the radius of Earth and 318 times its mass. Juno wasn't a typical NASA mission—it was designed specifically with the interior of Jupiter in mind; scientists wanted to investigate the planet's core to discover how the interior manifests its incredible magnetic fields and radiation belts, which bathe the innermost of its 92 moons—especially Io, Europa, and Ganymede, but also smaller, lesser-known moons like Almathea, Metis, Thebe, and Adrastea—in doses of radiation that would be deadly to any would-be explorers.

This meant that *Juno*'s engineers had to build a "radiation vault" to protect the suite of instruments, including magnetometers, plasma and particle detectors, microwave radiometers, and more. The craft was initially designed for a near-circular orbit a few thousand miles above the giant planet's cloud tops. At this close orbit, the camera would have fried within a few trips around the massive world—which mission engineers knew going in. But an attempt to shorten its orbit failed, leaving the craft in a 53-day orbit that swings it close to Jupiter but, mercifully, takes it out of the worst of the radiation as it leaves the close pass of the planet, which extends the life of the craft and leaves its camera online.

Hao Cao, a UCLA planetary scientist, has worked on both the Juno mission and the now-defunct Cassini mission to Saturn, which launched in 1997 and explored Saturn's system from its arrival in 2004 to its intentional plunge into Saturn in 2017. Although there are a lot of differences between the two planets, they're more like each other—hydrogen and helium surrounding dense planetary cores—than they are any other planets. Cao's research focuses on the interiors of planets, as well as their magnetic fields and dynamos. A few unusual things have come about in the time he's studied both worlds.

Some of the interesting features about Jupiter and Saturn's interior and radiation environment are apparent on their surfaces. For instance, there are significant winds on both planets, with depths of several thousand kilometers. "To understand the magnetic field of giant planets, we also need to understand their deep winds," Cao says.

On Jupiter, these winds can whip around at 250 to 330 feet per second at the cloud tops and stretch down to about 1,800 miles below the cloud tops, or 4% of that planet's radius. But on Saturn, that can be over 1,000 miles per hour and as deep as 5,500 miles, or about 15% of the radius. But where Jupiter is known for violent radiation, Saturn is comparatively more sedate—less than half as strong despite similar interiors.

Cao says many scientists studying these winds feel that they either form like winds do on Earth—through moist convection in the atmosphere—or that they're caused by convection rising from deep within the planets. Research published by a group of Juno mission scientists that included Cao in 2019 suggests they can also shear the magnetic field, creating variations in the field detectable by Juno. They may go deep enough to where atmospheric pressures turn gases into a form more akin to liquid metal.

But one of the most startling revelations of the missions has to do with the planets' cores. They're "diffuse" cores, with the rocky and gaseous material mixed in near the center—not quite what we're used to with a planet like Earth, where there is a sharper dividing line between the rocky mantle and iron core. "Essentially, we don't know if any parts of these planets are solid," Cao says. In fact, the core of Saturn could be distributed to up to 60% of the total radius of the planet, while Jupiter's may take up 30%–50%.

Saturn hasn't had its own mission version of Juno—not yet, anyway—but Cao hopes that any future mission would have an atmospheric probe to study the inside of the planet much like a 1995 probe of Jupiter aboard the Galileo mission that descended 125 miles into Jupiter's atmosphere before interior pressures and temperatures destroyed it. He also hopes for a microwave radiometer, which can penetrate below the cloud layers to characterize the interiors of the gas giants, and build out a better understanding of Saturn—including why it is comparatively more sedate than Jupiter.

FURTHER READING

Crockett, Christopher. "Jupiter Revealed." *Knowable Magazine,* June 5, 2020. https://knowablemagazine.org/article/physical-world/2020/what-has-juno-learned-about-jupiter.

European Space Agency. "Saturn's Magnetosphere." Accessed February 28, 2023. https://www.esa.int/Science_Exploration/Space_Science/Cassini-Huygens/Saturn_s_magnetosphere.

NASA. "NASA's Juno Finds Changes in Jupiter's Magnetic Field." News release, May 20, 2019, https://solarsystem.nasa.gov/news/947/nasas-juno-finds-changes-in-jupiters-magnetic-field/.

NASA Cassini Mission. "Magnetosphere." September 4, 2018. https://solarsystem.nasa.gov/missions/cassini/science/magnetosphere/.

Witze, Alexandra. "Jupiter's Stormy Winds Churn Deep into the Planet." *Nature*, October 26, 2017, https://www.nature.com/articles/nature.2017.22866.

THE MANY PHASES OF WATER

―――――――

As a graduate student, my main research topic was the unusual magnetic fields of Uranus and Neptune. Because a planet's magnetic field is generated in its interior, it's likely that there's some connection between the unusual magnetic fields and properties of these planets' interiors. Uranus and Neptune are known as the "ice giant" planets because they're mainly composed of ices like water, ammonia, and methane. But high pressures and temperatures inside the ice giants mean that these materials behave very differently than what we experience in our daily lives here on Earth's surface.

On Earth, we're used to experiencing water in a variety of phases. The humidity on a hot summer day, for example, is due to water vapor; the gaseous phase of H_2O. Lakes and oceans are full of the liquid phase. We experience water in its solid phase as snow and ice. But the solid water ice we experience here on Earth is only one of 19 known solid phases! It's called "normal hexagonal crystalline ice" and labeled ice I_h (pronounced "ice one h"). In this phase, the main crystal structure involves water molecules arranged in groups of six, with the six oxygen atoms in the water bonded to each other through sharing a hydrogen atom (known as hydrogen bonding) and forming a hexagon. I think of six people in a circle holding hands as a visual representation of this phase.

But this crystal structure is not the only possibility for solid water ice. For a range of pressures up to about 1,000 bars, if the temperature drops below about −150 degrees Fahrenheit, ice crystals transform into Ice I_c (ice one c), a cubic structure rather than hexagonal, and at an even colder temperature of −330 degrees Fahrenheit, ice is found in phase Ice XI (ice eleven) that looks like a three-dimensional rectangle. And yes, there is an Ice IX (ice nine) although it doesn't have the sinister properties imagined in Kurt Vonnegut's iconic novel *Cat's Cradle*.

Admittedly, the coldest temperature ever recorded on Earth is −129 degrees Fahrenheit, so we don't see Ice I_c or Ice XI here at the surface, but these temperatures can be found on the icy moons of the outer solar system, like on Europa, Enceladus, and Triton, and on the dwarf planets Pluto and Charon. And at even higher pressures and temperatures, one can find some of the other solid phases.

But what about the insides of Uranus and Neptune? Here the pressures can reach hundreds of thousands to millions of bars, and the temperatures are thousands to tens of thousands of degrees. What is water like in these conditions?

In the atmospheres of Uranus and Neptune, there are small amounts of water vapor mixed in with the hydrogen and helium. But as we go deeper in the planets, the ices, like water, ammonia, and methane, become the main composition, with water being the major molecule. At depth, the pressure increases, and so the water molecules come closer together and they become hotter. Eventually they will be close enough to be attracted to each other by intermolecular forces (these are essentially the electric forces felt between charges in the molecules). They're then no longer in the gaseous phase and behave more like a liquid.

As we descend further into the planet, pressures and temperatures get even higher, so new phases of fluid water are encountered. The first is "ionic water" in which the hydrogen and oxygen in the water molecules separate, creating hydrogen ions and oxygen ions. This soup of ions has the bulk elemental composition of water (i.e., two hydrogens for each oxygen), but these elements don't appear together in distinct molecules.

The best thing about ionic water is that it's electrically conductive because of those ions, that is, it can carry currents. It's in this ionic water layer in Uranus and Neptune that convection generates these planets' magnetic fields, and recall, those are weird. By weird I mean that the magnetic fields of Uranus and Neptune aren't mainly dipolar, with one north magnetic pole and one south magnetic pole, like we have at Earth and the other planets with dynamos. The magnetic fields are more complex and have several north and south magnetic poles, in various locations on the surface. We call this type of field a "multipolar" field.

Why are they like this? Well, that's what I wanted to understand during my PhD studies. Since Uranus and Neptune are the only planets with multipolar fields, and they're the only planets with dynamos generated in ionic ice layers, it was worth testing whether there was a connection. Previous researchers had suggested that only the thin outer layers of these ionic regions were actually convecting. That would mean that the dynamo is only being generated in this outer thin shell. The thickness of the convecting layer matters because it determines the length scale of the fluid motions in the layer. In a thin layer, you can't have large wavelengths of motion; there's just no room for them.

So there I was in graduate school, tackling the problem using computer models. We can take the equations that govern fluid motions and the equations that govern magnetic field generation and program a computer to use these equations to simulate what is happening in a planetary dynamo. These models are similar to climate models in that they determine what the fluid motions and temperature fields are like over time, based on the fundamental equations (which are too complex to solve with pen and paper). And just as climate models can predict what the winds and ocean currents will look like given certain conditions, dynamo models can predict what the magnetic fields will look like given certain conditions.

For Uranus and Neptune, the condition I was testing was how thin this convecting ionic layer is. Most dynamo models that had been run at that time were focused on Earth's magnetic field. There, the geometry of the Earth's outer core was used in the models: you input a solid inner sphere to represent the solid iron inner core and surround it with a liquid iron shell that's about two-thirds the radius of the whole core. It's in the liquid shell that convection is happening and creating Earth's magnetic field. For Uranus and Neptune, the geometry was different. I input a very large inner sphere (in this case a fluid sphere but not convecting) and surrounded it with a thin convecting shell where the dynamo would be produced. After running the models (and fixing many, many, many bugs in my coding) I was ecstatic to find that this scenario produced multipolar fields, just like in Uranus and Neptune. So, I now had a plausible answer for why Uranus's and Neptune's magnetic fields were so weird.

I remember writing my thesis on the subject and having to answer the question I heard many times from fellow scientists: "But if the ionic layer is thin, then what is below this layer?" That is: why is the inner sphere *not* convecting? At the time, I didn't have a good answer; actually no one did. It's extremely challenging to carry out experiments and computer simulations of the chemistry of water at the pressure and temperature conditions of Uranus's and Neptune's interiors, so we didn't really know what happens to water under this ionic layer. The best answer I could give was: "We don't know, but something must be different in the inner half of the planet that makes it not participate in dynamo action."

There were possibilities suggested to explain this. We really didn't know much about the composition of the ice giant interiors. Maybe the other ices, like ammonia and methane, that are mixed in with water do something funky at these higher pressures. Maybe rocky materials mix in with the water at deeper depths, and that somehow makes the material more viscous, or less conductive, or not convect. Maybe Uranus's and Neptune's cores are fuzzy the way we now know Jupiter's and Saturn's core are likely to be. Or maybe (and admittedly this is my favorite), a new phase of water occurs with different material properties. There were hints of this at the time of my PhD in experiments and theories about a new phase of water at high pressures and temperatures, dubbed "superionic water."*

One of the most satisfying moments I ever experienced in research occurred in 2011, when I downloaded a paper by another research team that had used computer simulations of quantum chemistry to demonstrate that yes, superionic water would be the phase below ionic water in the ice giants, and it could be stable to convection.[6] I suddenly had a possible answer to that question which had vexed me for years and my theory for the cause of Uranus's and Neptune's unusual magnetic fields appeared to be more solidly grounded.

---

*I've noticed an alarming trend in physics to add the prefix "super" to make cool things sound even cooler. Examples include superconductor, super-Earth, supermassive black hole, and superstring theory. It would be interesting to see if there was any relationship to increased funding on a topic when the prefix "super" is added. If so, I propose a new field to be known as superdynamo theory.

Then in 2019, experiments at the Omega Laser Facility in Rochester, New York (involving lasers that create shock waves that heat and compress materials to temperatures and pressures found in ice giants' interiors), discovered the crystal structure of superionic water: The oxygen ions join together to create a crystal lattice, while the hydrogen ions (i.e., protons) diffuse through the lattice, sort of swimming around.[7] The freedom of the protons to move around is what makes superionic water so electrically conductive, and the crystal structure of the oxygen lattice makes it a solid.

Although computer simulations of quantum chemistry had previously demonstrated superionic water *should* exist, experiments are the real thing. They demonstrate superionic water *does* exist. This is why scientific experiments are so important—the proof is in the pudding, as the saying goes.

An open question at the moment that my research group is working on is: "How solid is superionic water?" By this I mean: what is the viscosity of superionic water, or how easily can it deform and flow? Its viscosity would determine whether it convects (perhaps at a much slower rate) and whether it can participate in the generation of Uranus's and Neptune's magnetic fields. If it's very viscous, it won't, and the model I proposed in graduate school to explain their magnetic fields might hold water.*

---

*Pun intended.

HIDDEN WATER OCEANS

Exotic phases of water are very exciting, but water also appears in other places in our solar system in unusual ways. We're particularly hell-bent to find where it occurs as a regular liquid. That's because our understanding of how life forms suggests that liquid water is a necessary ingredient for the creation, or at least the persistence, of life. Water acts as a solvent to help break bonds between certain molecules, allowing them to create longer chains of molecules, eventually long enough to be as complex as the building blocks of life, like amino acids. Our search for life outside of Earth has therefore led to a search for liquid water—and we're finding it in some unexpected places—for example, as vast global oceans on tiny frigid moons in the outer solar system.

In the 1990s the Galileo mission visited Jupiter and its family of moons. The four biggest moons—Io, Europa, Ganymede, and Callisto—are almost like their own solar system, orbiting Jupiter at the center. Each moon is a unique world. Io is covered in active volcanoes, a result of the tidal forces stretching and heating its insides and leaving the moon bone dry, removing any water it had previously. The other three moons are covered in miles of solid water ice; the Galileo mission discovered that under this ice in these bodies are global shells of liquid water.

Perhaps the most exciting of these three ice-covered moons is Europa. It's the tiniest of the Galilean moons, with a radius of just under 1,000 miles. That means it's smaller in diameter than Earth's moon and even than the span of the continental United States. But in this tiny moon, the subsurface ocean has more than twice the water in all of Earth's oceans combined. And importantly, Europa's subsurface ocean likely has other necessary ingredients for life, such as energy sources and organic molecules. The Europa Clipper mission is scheduled to launch in 2024 to try to sample some of that water, not from the depths of the ocean, but from giant geysers that spray ocean materials through cracks in the surface and out into space. (On Earth, geysers like those at Yellowstone are powered by areas where water comes into contact with hot rocks or magma, whereas geysers on Europa are powered by tidal forces from Jupiter.) This mission has the potential to answer the question that has fascinated humans for millennia: is there life beyond Earth—and if so, in what form?

The subsurface oceans on Europa, Ganymede, and Callisto weren't discovered by flying a spacecraft through geysers; they were detected in a more indirect, but still very cool, way—through their interactions with Jupiter's magnetic field. This was possible because the oceans aren't "pure" water but likely have some amount of salts in them. These salts, just like salt water on Earth, make the oceans a bit conductive, so they can transport electrical currents. Not enough currents to generate a dynamo, but enough to create a magnetic signature that can be detected by spacecraft.

I love this story of how the Galilean moons' oceans were discovered because it highlights that not only might magnetic fields be important for protecting life (like it is on Earth) but they also might be important for detecting habitable conditions for life (i.e., the presence of liquid water).

The Galilean moons aren't the only outer solar system bodies with subsurface oceans. Consider one of my favorite bodies in the solar system:* Saturn's moon Titan. Like Ganymede, Titan is bigger than Mercury—although Mercury has more mass than both worlds combined owing to its incredible density. Titan also has very little iron and likely no iron core at its center. In addition, the interior isn't fully differentiated—the rocky material is likely mixed in with some ice throughout Titan's deep interior.[8] But above that deep interior, Titan has a global subterranean water ocean.

There are some lakes, rivers, and seas on Titan's surface as well, but they aren't made of water, and they won't be resort areas anytime soon—they're made up of liquid hydrocarbons like methane and ethane. Titan is the only body in our Solar System other than Earth to have standing liquid on its surface—just not water. Because Titan has a subterranean ocean, some scientists believe there may be a chance for life below its surface. But the lakes of hydrocarbons on the surface could have

---

*Full disclosure: I call all the planetary bodies in the solar system my "favorite" at one point or another. Except Venus.

life, too—a big maybe. A 2015 study found it was possible to build a cell-like wall out of chemicals available on Titan, using acrylonitrile (a type of vinyl) in place of the lipid layer we're used to.[9] But of course, that's wild speculation.

Voyager also had a lot to tell us about another moon of Saturn that Cassini scientists were excited to visit—Enceladus. This moon is tiny, just 300 miles across, meaning if one were to pluck Enceladus out of the sky and place it on the surface of the Earth, it would just cover the distance from Philadelphia to Pittsburgh. Enceladus is mainly made of water ice and has a few craters on its surface. Its south pole is somewhat unusual, though, covered in what scientists have nicknamed "tiger stripes"—parallel fracture lines on the south pole of the moon.

Those tiger stripes have water from deep below Enceladus shooting out of them to more than a hundred miles into space,[10] in one of the entire solar system's most impressive acts of cryovolcanism.

Enceladus has a truly remarkable interior. Its ice shell is about twenty miles thick, with a six-mile-deep global ocean underneath it. Then there's a rocky center under that, with the possibility of direct contact between interior hot rocks and the ocean floor. Despite repeat visits by the Cassini mission, we're still figuring out what that interior looks like, but one model holds that Enceladus's core may actually be porous—up to 30% empty space into which water can enter, much like a sponge. Water can worm its way deep down through the rocky center and come in direct contact with materials experiencing radioactive decay at the center. This is similar to how hydrothermal vents in our own ocean floor come into direct contact with ocean water, where life like the ironclad sea pangolin snail in the Indian Ocean can thrive (at 750 degrees Fahrenheit among toxic subterranean sulfide spews, which spits in the face of what we consider habitable environments). Cassini also discovered evidence that Mimas, an Enceladus-sized Saturnian moon that has one huge crater making it look exactly like the Death Star from Star Wars, may have an ocean as well,[11] although there are no signs of a spectacular geyser show like at Enceladus.

Even further out in the solar system are Triton, which is the largest moon of Neptune, and Pluto, the largest known object in the Kuiper Belt, which is a region of small, icy bodies past the orbit of Neptune that stretches out about 1,000 times the Sun-Earth distance deep into the solar system. Triton orbits Neptune backward, (i.e., in the opposite direction of Neptune's rotation) which leads planetary scientists to believe it was formed in the Kuiper Belt and then captured by Neptune later. Triton has an interior core of rocky silicates covered by high-pressure ices, then an ocean, then a crust of water and nitrogen ices. It has evidence of cryovolcanism too, discovered by the Voyager 2 mission in 1989. Since we haven't returned in the interim, their nature remains a mystery, but they're likely due to interactions with Neptune pulling up material from a subsurface ocean.

The potential oceans below Pluto's surface are likely kept liquid through the presence of ammonia acting as an antifreeze and Pluto's proximity to its moon Charon, if you can even consider it a "moon"—Charon is half the size of Pluto, and thus Charon doesn't even orbit Pluto—it's better described as Pluto and Charon both orbiting a common center of gravity well outside Pluto. The tidal interactions between the bodies may provide a way for Pluto to have a possible liquid water ocean underneath.

Even the dwarf planet Ceres, located in the asteroid belt between Mars and Jupiter, has evidence for pockets of oceans below its surface. It seems like everywhere we look in the outer solar system, we find more oceans. I picture Oprah coming out on a stage with an audience full of outer solar system bodies and she excitedly points to each of them: "You get an ocean, and you get an ocean, and you get an ocean. . . ."

The solar system appears to be littered with moons that have global subsurface oceans. Earth may be somewhat unusual, then, by hosting its oceans on the surface of the planet.

CARBON: FROM GREENHOUSE GAS TO DIAMOND ICEBERGS TO DIAMOND EXOPLANETS

I think about carbon a lot. It exists as part of the greenhouse gas, carbon dioxide, in Earth's atmosphere, and it's crucial to making Earth a habitable planet. First, it provides the energy source for plants that, through photosynthesis, create the oxygen we humans and other animals breathe. And as a greenhouse gas, carbon dioxide in the atmosphere has warmed Earth's surface temperature about 60 degrees Fahrenheit compared with what it would be without any greenhouse gases.

This is the "good" side of the greenhouse effect, and through this natural process, Earth's surface is at the right temperature for water to exist as a liquid. Since liquid water is so crucial to life as we know it, we think of liquid water as a necessary ingredient for life to exist. Indeed, the search for life on other planets, as well as the search for exoplanets that might be habitable, is based on criteria involving liquid water on the surface.

Unfortunately, humans are disrupting the natural balance of carbon dioxide in the atmosphere, primarily through the burning of fossil fuels. This is causing excess carbon dioxide in the atmosphere and is responsible for global climate change. We're already experiencing the consequences of global climate change today, like more devastating storms and wildfires. Global sea levels are also rising, and the chemical balance in the oceans is changing, threatening ocean life populations. Empirical data and models indicate these consequences will get worse if we don't reverse current warming trends.

Carbon dioxide is also present in other planets' atmospheres, where it's the main constituent of Venus's thick atmosphere and Mars's thin atmosphere. The carbon we have in the atmosphere also sometimes ends up deep inside the planets. As we saw in chapter 3, the diamond phase of carbon is important in bringing samples from the deep Earth up to the surface. Other planets likely have diamonds, too. And sometimes they can demonstrate some unusual behavior.

Consider Uranus and Neptune. It turns out diamonds can "rain" in Uranus and Neptune, just like helium rains in Jupiter and Saturn. Carbon, locked up in the molecule methane (CH_4), is present inside Uranus and Neptune. It's methane gas in the atmosphere (yes, Uranus smells like farts; every planetary scientist has heard this joke, so please bring new material). As we go deeper into the planet, eventually the pressure and temperature cause the molecular bonds in methane to break apart, and carbon is freed from the hydrogen enabling it to form diamond crystals. Although we haven't sampled these Uranus/Neptune diamonds directly, high-pressure experiments have demonstrated their existence. For instance, a 2022 experiment at the SLAC National Accelerator Laboratory recreated the diamond rain in conditions that mimicked the ice giants, firing lasers into a dense plastic to produce them.[12] Firing a laser into food-grade plastic may seem strange, but it strikes a good balance between carbon, hydrogen, and oxygen like you might see that deep down.[13]

Deep in those planets, the temperature gets high enough that the diamond melts. Experiments have shown that the melted diamond is denser than the surrounding materials and so sinks to the bottom. This suggests that there might be seas of liquid diamond deep inside Uranus and Neptune.[14] (It's probably more accurate to say liquid carbon, but liquid diamond sounds cooler.) And these seas might even have diamond icebergs! That's because carbon shares the same unusual property that water does near the freezing point: frozen diamond is slightly less dense than liquid diamond, and so it floats, just like icebergs do on the ocean.

The amount of carbon (and hence diamonds) in our solar system was determined by the initial fraction of carbon in the solar nebula from which the Sun and planets formed. The relative abundance of several elements in the solar nebula is primarily responsible for what minerals form. For example, rocks on the terrestrial planets are made mostly of iron and magnesium mixed with silicates. Those silicates are created from the element silicon bonded to oxygen atoms. This bonding happened because of relative abundances of oxygen and carbon in our solar nebula. But what if another star's nebula has different relative abundances?

By examining light coming from other stars, we can determine their composition. Different elements absorb and reflect in different wavelengths (or colors), and a spectrograph can look for these colors to identify a chemical coming from a star. As we've seen with our solar system, the composition of the star is strongly related to the composition of the building blocks of planets (remember the chondrite meteorite/Sun composition comparison in chapter 3). It turns out we've measured different compositions for other stars. The ratio of carbon to oxygen in our solar system is about two-to-one (so there are almost two oxygen atoms for every carbon atom out there).[15] But if we look at other stars in our neighborhood, and focus on ones that are similar to the Sun (i.e., same size, age, and temperature), we find that about 10%–15% of those stars have larger carbon-to-oxygen ratios.[16] Planets in these systems have more carbon in their building blocks, and that could affect the types of minerals that form there. Silicon could bond more easily with carbon to create silicon carbide than with oxygen.

Scientists have theorized what the interior structures of these "carbon planets" might look like.[17] Planets with masses similar to that of Earth might have shells of graphite (like in pencil lead), surrounding thick shells of diamond or silicon carbide. The surfaces of carbon planets would lack water,[18] making them inhospitable to life as we currently understand it.

Indeed, we may have already discovered one of these carbon planets: 55 Cancri e is an exoplanet whose mass is eight times that of Earth, and its radius is twice that of Earth. Based on what we know of this planet's star, it's possible that it's rich in carbon.[19] The planet is about 40 light years away from Earth, so before planning your travels there to hoard diamonds (or pencils, if that's your thing), keep in mind that it would take

40 years to travel there at the speed of light. Also, the planet would probably kill you since the surface is so hot that it's completely molten, which makes mining difficult. Maybe not worth the trip.

UNUSUAL FILLINGS FOR VOLCANOES

The final unusual planetary composition we'll consider in this chapter is the composition of volcanoes. Our experiences with volcanoes on Earth are limited to the basaltic magma-spewing rock volcanoes at hot spots over mantle plumes like at Hawaii or at volcanoes in ocean island arcs like Mount Fuji, caused by the release of volatile gases as a lithospheric plate descends into the Earth at a subduction zone. We've already discussed cryo-volcanism, where liquid water or similar chemicals are pushed out of cracks in solid ice. But there are also mud volcanoes that form when hot water interacts with rocks below ground and depressurizes through a crack or fault in the Earth's surface. And of course, there are the table-top volcano experiments you can do by mixing appropriate amounts of vinegar and baking soda (and scary-looking food coloring, obviously) to create the release of carbon dioxide gas, causing fizzing liquid to spill over a playdough volcano mold.

All the rocky planets have evidence of rock volcanic activity at some point in their history. Venus may even have active volcanoes today, and Jupiter's moon Io is definitely volcanically active, as we can watch volcanoes launch plumes of lava out onto its surface in real time.

In addition to "rock" and "ice" volcanoes, the future Psyche mission to the asteroid 16 Psyche might also find evidence of a new type of volcano: a metal volcano.[20] 16 Psyche is believed to be mostly made of iron. As the surface of 16 Psyche cooled, it's possible that ferrovolcanism (i.e., with lava of iron) resulted. We've never seen a metal volcano before, and the Psyche mission may offer us the first glimpse of what they look like (for more on the Psyche mission, see pages 78–81).

Volcanoes of Fire and Ice

The sight of magma bursting up from miles inside Earth then streaming down a volcano's face as glowing orange lava is an enthralling and beautiful expression of our planet's powerful internal forces. The inspiration for legends including those connected to Hawaii's goddess Pele, and the source of tragedies including the destruction of Pompei in 79 CE, volcanoes today provide critical information about the materials and movements that lie both miles below our feet and beneath the surfaces of other celestial bodies.

Earth hosts a variety of volcano types—including the familiar examples (cinder, composite, and shield) and more exotic phenomena. We generally think of these as explosive mountains that release lava—a combination of oxygen, silicon, aluminum, iron, calcium, sodium, magnesium, and potassium. But there are also volcanic vents at the bottom of the ocean and under lakes that slowly seep volatile gases like methane or carbon dioxide into the lowest layers of the water, with sometimes deadly consequences, as happened with the limnic eruption in Cameroon's Lake Nyos in 1986. Geysers like those at Yellowstone National Park are also areas where cooling volcanic rocks from below-ground magma heat chambers of subterranean water, which explode when pressures below reach too great a level. The US Geological Survey estimates that there are about 1,350 potentially active above-ground volcanoes worldwide, many of which are located in the Pacific Rim's "Ring of Fire."

Earth isn't the only home to volcanoes. Venus, for instance, has lightning storms (caused when ash particles collide, generating static electricity in the erupting plume) and resurfacing events that gesture at present-day volcanism—something future Venus missions may be able to verify. A study in 2023 of old Magellan data even discovered evidence of a volcanic vent changing shape on Venus over time, indicating some kind of volcanic activity. Mars once had active volcanoes as well, although they've since gone quiet. There are also cryovolcanoes on several outer solar system worlds covered in this book, the first of which was discovered in 1989 by the *Voyager 2* probe on the surface of Neptune's moon Triton.

But Lynnae Quick, a planetary geophysicist at NASA's Goddard Space Flight Center, says there's a celestial body close to us with present-day volcano-like activity—but maybe not as we know it. The Moon was once volcanically active; there may be some magma still moving beneath its surface. This is called intrusive volcanism, whereas volcanoes that erupt at the surface are extrusive.

"We don't have active volcanism as far as lava spilling out onto the surface, but there are regions of Earth's moon where we do believe that we have silicate melt, that's kind of moving around beneath the surface," Quick says. Similar underground movements may be taking place on Mars.

But that's not all. While cryovolcanism at various icy worlds is discussed in chapter 7, there are some other strange volcanoes to look at. Quick worked on the Dawn mission to Ceres, the largest object in the asteroid belt and probably a vestigial protoplanet. A detection of water vapor by the Herschel Space Observatory left planetary scientists speculating that it could have geysers. Quick says the team didn't see that, but there was evidence for other kinds of volcanism.

As the craft approached Ceres, it saw bright spots sticking out from various craters. "What it looked like was there had been some cryovolcanism on Ceres, and it had formed these bright spots," Quick says. As the Dawn team investigated, they found that it likely was a "briny cryomagma" from ancient flows. The bright spots are leftover salts once the water part of the cryomagma sublimated away—that is, turned from a solid directly into a vapor—thanks to being much closer to the Sun than the icy worlds of the outer solar system.

But the team also found a surface feature called Ahuna Mons, which they called a "lonely cryovolcano," as it's the only major mountain across the surface of Ceres. It's a relatively young feature, and Quick says its flows were likely a mix of briny water and mud made of sodium carbonates, making it a mud volcano. Over time, other similar volcanoes probably "relaxed away" once the water ice sublimated.

More atypical cryovolcanism could exist on a couple other worlds in the solar system. Ariel and Miranda, two moons of Uranus, may have what Quick calls "very thick, very viscous, very sticky cryolava flows" from the presence of ammonia and methanol. "The inclusion of methanol would make for a very viscous fluid when combined with water and ammonia," Quick says.

Magma, mud, ice, and briny water aren't the only materials that volcanic forces bring to the surface; on Earth, diamonds in kimberlite and black-green obsidian, formed from melted and cooled silica after eruptions, are found as well.

FURTHER READING

NASA. "NASA Discovers 'Lonely Mountain' on Ceres Likely a Salty-Mud Cryovolcano."
  News release, September 1, 2016. https://www.nasa.gov/feature/goddard/2016/ceres
  -cryo-volcano.

NASA Jet Propulsion Laboratory / California Institute of Technology. "NASA's Magellan
  Data Reveals Volcanic Activity on Venus." News release, March 15, 2023. https://www
  .jpl.nasa.gov/news/nasas-magellan-data-reveals-volcanic-activity-on-venus.

Smithsonian Institution, National Museum of Natural History, Global Volcanism Program.
  "Worldwide Holocene Volcano and Eruption Information." Accessed February 28,
  2023. https://volcano.si.edu/.

Thornton, Stuart. "Cold Explosion." *National Geographic,* September 27, 2022.
  https://education.nationalgeographic.org/resource/cold-explosion/.

CHAPTER 7

The Future of Planetary Exploration

ASTRONOMY ISN'T JUST ABOUT LOOKING to the skies and designing new missions. Sometimes it involves studying the work of scientists who came before us, and working in labs in myriad ways to collect volumes and volumes of data, a lot of which can be referenced at a later time to find, or better understand, new discoveries. For instance, Pluto was officially discovered in 1930, but astronomers digging through archives since then have found evidence of its discovery going farther back, at least to 1914,[1] and possibly to 1909.[2] This is called *precovery*, as distinct from *discovery*.

How'd they miss it? Simple—they weren't looking for it. It was on old photographic plates—exposures on glass, intended to track the night sky and the positions of stars and planets. Galileo observed Neptune back in the seventeenth century, but he didn't know what he was looking at. Astronomers didn't officially find Neptune until they were observing it due to strange motions from Uranus's orbit hinted at the pull of a more distant object. When the Carnegie Observatories went back through some old data on a white dwarf in 2015—the kind of zombie star our Sun will one day become—they were surprised to find something strange in data from 1917.

As mentioned earlier, exoplanets went undiscovered until the 1990s, after decades of false starts. There weren't *really* any examples of precovery for exoplanets, because the techniques used to discover them didn't come into their own until the 1980s.

But there were a few sprinklings of evidence before then that exoplanets *could* form, largely from the work of the Infrared Astronomical Satellite (IRAS), a NASA space telescope from the 1980s. It observed debris disks—those sorts of leftovers from star formation—around Epsilon Eridani and other stars, although at the time the evidence required confirmation by subsequent observations.

Walter Adams, former director of the Carnegie Observatories, gathered the 1917 spectra data at the Mount Wilson Observatory. Adams looked at an odd celestial object called van Maanen's Star, which is about fourteen light years away. Astronomers in the early twentieth century had taken spectra of van Maanen's Star not knowing what they were looking at on a few levels. For instance, Adriaan van Maanen didn't know he was looking at a white dwarf—he assumed it to be an ordinary star upon its discovery that year. It was another six years before

Willem Jacob Luyten would coin the term "white dwarf" and correctly classify van Maanen's Star as part of this new class of star, although at the time astronomers didn't understand what they were, only that they appeared whiter than they should given their small size.[3]

The spectra of the atmosphere of van Maanen's Star showed that it was made of helium and also had unexpected amounts of calcium and iron. It's unexpected because any element heavier than helium should sink into the deeper layers of the white dwarfs and therefore not be observable in the atmosphere. More recent measurements of many white dwarfs suggest that about 3.5% of them have these unexpected amounts of heavier elements such as calcium and iron, but where do they come from?

One intriguing theory is that they form from fragments of planets and asteroids that are pulled into the star, polluting its surface. These white dwarfs may have had whole planetary systems while they were regular "living" stars. But while the stars died, so did the planets. It therefore seems that planetary fragments or asteroids were falling down to van Maanen's Star when it was observed in 1917, and that means we've had evidence of exoplanets since 1917, even if we didn't know it.

Polluted white dwarfs are particularly fascinating because in some ways, we're seeing the insides of these exoplanetary bodies in the atmospheres of these stars. It turns out it may take a dying planet to really get a sense of things.

THE DEATH OF A PLANET

Stars can be quite dramatic—giant balls of nuclear burning, giving off immense light, heat, and high-energy particles. Depending on how massive the original star was, upon its death it might produce a supernova explosion, flinging a shockwave of material from the star's outer layers into space. This is triggered when the star is no longer able to fuse hydrogen into helium, and the heat from the core of a star can no longer keep up with the pressure of gravity from the mass of the star. If they're massive enough, dying stars may even produce balls of pure neutrons, or even black holes.

But what about planets—do they die, too? We've discussed how they're "born" in chapter 2 during the formation of the solar system. But is there any equivalent to the other end of a planet's life cycle? Here are two possibilities.

First, you might consider a planet dead when it runs out of heat. Planets are born hot, with the processes that create them storing heat in their interiors. They also have sources of heat from radioactive decay and some chemical processes. But these heat sources will eventually wither away, at which point the planet will cool until it reaches the temperature of the space surrounding it. At this point, we could expect the interiors of the planets to be motionless, with no convection around to drive convection or magnetic field generation.

That point in time is extremely distant in Earth's future. As a quick example, the most abundant radioactive element in

Earth today, uranium-238, has a half-life of 4.5 billion years. That essentially means we've only used up about one-half of our uranium reserves since the solar system began. And there are other radioactive elements in Earth (although not as plentiful, so they don't produce as much heat) with even longer half-lives, like rhenium-187 and samarium-147 with half-lives of 44 billion years and 106 billion years, respectively.

The other possibility is actually the more realistic one for Earth. Our planet will meet its untimely demise long before it runs out of heat. Instead, the planet's fate is strongly tied to that of the Sun. Although our Sun is too small a star to create a supernova when it stops nuclear burning, it will go through a phase of life known as the "red giant" phase. After that, it will become a white dwarf.

Hydrogen in the Sun's core will eventually run out as stellar fusion creates helium from the hydrogen. This will happen about 5 billion years from now. The Sun will then contract, and this will cause hydrogen to begin fusion in an outer shell of the Sun. This will add more heat to the Sun and cause the Sun's atmosphere to start expanding outwards, which isn't good news for the planets. Ballooning outward, the Sun will eventually absorb Mercury, Venus, and likely Earth during this time.[4] In this scenario, Earth won't last forever, although it will last quite a long time.

Alien astronomers watching the Sun after that happens might be able to see a sudden pollution of heavy elements into the white dwarf version of our future Sun, much like our own astronomers were able to witness with van Maanen's Star. We've seen it happen with other white dwarfs, like G238-44, where astronomers found nitrogen, oxygen, magnesium, silicon, and iron in the spectra.[5] Oxygen and iron show up in stellar atmospheres, but the exceptional abundance of iron indicated that it was coming from planetary material. And the abundance of nitrogen, which also shows up in stellar atmospheres in small amounts, indicated that icy material was falling into that star, too.

The solution had to be planetary debris falling into G238-44. A UCLA undergraduate, Ted Johnson, told NASA in 2022 that "the best fit for our data was a nearly two-to-one mix of Mercury-like material and comet-like material, which is made up of ice and dust. Iron metal and nitrogen ice each suggest wildly different conditions of planetary formation. There is no known solar system object with so much of both."[6]

These kinds of observations aren't *just* limited to dead stars. In 2022, astronomers working on the TESS (Transiting Exoplanet Survey Satellite) mission discovered three peculiar Jupiter-like planets around main sequence stars. Their home stars are *dying,* but they aren't dead yet (or as Miracle Max in *The Princess Bride* would say, they're "just mostly dead"). That's still enough to pull the planets closer and closer in and, eventually, rip them apart, even before the star fully dies. Already, two of the planets are puffing up from their death spiral. One of those, TOI-4329b, would, according to the authors, "provide a favorable opportunity for the detection of water, carbon dioxide and carbon monoxide features in the atmosphere of a planet orbiting an evolved star, and could yield new information about planet formation and atmospheric evolution."[7] In other words, we could observe that planet as its innards get pulled into its home star and determine some of the chemicals in the deeper layers of the atmosphere.

Other exoplanets are in the process of losing their atmospheres as radiation from their stars strips them away; those that lack cores are denied the protections of magnetospheres. In 2014, astronomers found a "comet-like" tail trailing the planet GJ 436 b.[8] Its home star, Gliese 436, isn't really expected to die anytime soon. It is part of a class of small stars that can survive on the scale of *trillions* of years because their low mass makes them relatively stable. Thus, astronomers may be able to watch as the atmosphere of this Neptune-sized world is stripped, leaving behind a core of materials that typically are only produced under intense pressure. We *think* we've seen what becomes of these planets through two very strange super-Earths, CoRoT-7b and Kepler-10b. Modeling of how much atmosphere planets like Jupiter and Saturn could lose if they came too near their home stars leads to planets a whole lot like those two at around the right distance from their stars. Astronomers believe CoRoT-7b has a surface with oceans of lava and we could, with the right telescope, observe whether it has an atmosphere that tells us more about its composition.[9]

These kinds of planets and planet fragments could provide us with our only glimpses into the insides of distant exoplanets. But to observe them in any detail, we'll need some pretty big telescopes.

THE FUTURE OF PLANETARY EXPLORATION

On Christmas morning 2021, if you woke up early, you could see the launch of NASA's new flagship telescope: the JWST. (Given its namesake's involvement in the Lavender Scares against LGBTQ NASA employees of the 1950s–1960s, I won't spell out the full name.)[10] The telescope then spent around six months carefully unfolding its mirrors, unfurling its sunshield, and calibrating its instruments, all so it could keep a keener eye on the universe than ever before.

JWST is huge. The primary mirror is about 21 feet across, whereas Hubble's is a little less than 8 feet. It's composed of 18 tiled, gold-plated individual hexagonal pieces. The sunshield is about the size of a tennis court to protect the telescope from the heat of direct sunlight, as the whole thing has to be kept frigid to work correctly. And that's because unlike our eyes, or even Hubble's "eyes," JWST can see only in infrared.

This may seem like an odd choice. If you have a telescope more powerful than Hubble, wouldn't you want it to have the same capabilities? You've probably seen infrared goggles in movies in the form of heat vision. Heat shows up in infrared, just below what our eyes can see. Infrared is good for a lot of things in space, as that part of the spectrum includes the light from objects that are very far away from us, and hence, redshifted to longer wavelengths. Hubble can view some near-infrared wavelengths, but JWST is built to go down into mid-infrared wavelengths. This makes it a crucial tool for staring at galaxies from a few hundred million years after the Big Bang.

But for objects nearer to us, visible light hasn't redshifted into the infrared. Instead, light in the infrared part of the spectrum is coming from the heat of an object. For instance, in fall 2022, JWST peered at the Pillars of Creation in the Eagle Nebula, the subject of an iconic Hubble photo. In these pillars, new stars are born, and with those stars come new planets. Like Earth, those planets are built of the stuff that their star was, not-so-lovingly assembled as dust became pebbles, which accumulated and differentiated. These objects got bigger and bigger, smashed into each other, underwent violent upheaval, and generally had some brutal pileups on the way to becoming a more orderly protoplanet system. But as a stellar nursery, these little stars are like watching what happened to our own home system 4.6 billion years ago, before there was even a microbe around to see it. With the JWST infrared images, we don't see the stars too much, since they don't emit a lot of infrared light, but we do see more of the gas and dust surrounding the stars. This gives us a glimpse of the birth of planetary systems and insight into what materials make up the regions surrounding stars. It's a stunning, evocative image.

But astronomers applying for time on JWST don't *just* want to stare at the nebula forging stars. They want to figure out what's going on in individual stars, too, whether in nebula or not. They want to compare these stars against the Sun, to figure out our own origins, as well as how the galaxy evolved over time.

Why is all this important? The world around us, as discussed from myriad perspectives in the previous chapters, is the result of a series of circumstances that made Earth a suitable place to evolve and maintain life. The eight known planets and hundreds of known moons and dwarf planets may all look quite different, but they're each just a rearrangement of the same fundamental base elements in similar abundances. Having enough iron is important to make a planet like Earth, and that iron is a cornerstone of our magnetic field. If you don't have enough, you could have a planet the same size as Earth and at the same distance from its star, but without the coveted source of protection our own planet has buried deep inside.

You can even think of it a bit like this, with, wait for it . . . another food analogy. If you've been in a kitchen enough, chances are you've had near misses with mixing up sugar and salt. Suddenly, your cookies are unpalatable, or your pretzel is an unintentional dessert. Both sugar and salt come in crystals. But imagine a world where instead of sugar and salt, you're dealing with other crystalline white powders like baking soda or strychnine. Your cookies are suddenly either *really* bad, or in the latter case, edible only once. But on the surface, cookies with these four ingredients still might look *exactly* like cookies.

JWST can't tell what's on the surface of a distant planet. It can't even tell us much about the planet at all. But it can do a few important things when it comes to exoplanets: it can directly image very young ones, and it can watch others pass in front of their stars and tell us about the one thing that could enable us to know whether we're looking at a sugar or a strychnine cookie: it can break down their atmospheres for certain chemical fingerprints.

There are other giant telescopes coming, like the Extremely Large Telescope (yes, that's the real name) under construction in Chile. These, and subsequent generations of telescopes, will be the tools we need to look at those distant exoplanets that have had haphazard lives and find materials that indicate what they're made of. They could inform us about planets that currently have a more well-rounded life—and tell us whether we're looking at a salt, sugar, or strychnine cookie.

Astronomers really want to find Earth-analog planets, and stars like TRAPPIST-1 and Proxima Centauri provide planets that are relatively easy to study. But they're also orbiting red dwarf stars, which produce violent flares. We need to look at more planets around these types of stars to figure out whether their magnetic fields are enough to hold onto an atmosphere, or whether their stars are going to overpower them, as may have happened on Mars. We can't gaze inside the planets themselves, but we can gaze at their atmospheres as they pass across. Having as much data as possible can tell us whether we can really find true Earth analogs out there.

Think of Venus, Earth, and Mars. All three are in the habitable zone of the Sun, but different histories led them to different paths. There's evidence that all three once had oceans, but something still poorly understood caused Venus to experience runaway climate change that turned this Earth-sized world into a sweltering, dry hell. NASA is working on a pair of missions to Venus that it hopes will fill in some of the gaps that led the planet to its present dire state. The European Space Agency (ESA) is working on a similar mission. As Venus may have once had oceans, these three missions could help us learn more about what conditions allowed life to persist and thrive on Earth.

We don't fully understand what happened to Mars's atmosphere, either. Is it too small for its early magnetic field to hold onto its atmosphere particularly well, or did the atmosphere get stripped away after the dynamo died about four billion years ago? Without an effective magnetic shield, solar winds could have stripped the planet of its dense atmosphere, drying up its lakes and rivers and dooming any life that was once there to, at best, live a subterranean existence . . . and that's an optimistic possibility.

Among JWST's first set of images—encompassing a deep field of distant objects, a quartet of strange galaxies, a brilliant supernova remnant, and a nearby star-forming region—was something that wasn't an image at all but was a way for NASA, ESA, and the Canadian Space Agency to show what their telescope could do. The telescope had looked at the transits of WASP-96b, a Jupiter-sized world in a tight orbit around its home star, and drew out the composition of its atmosphere by studying the spectrum of its light. We know that certain chemicals leave certain signatures in a planet's spectrum, and we could thus keep track of important things like water vapor.

JWST astronomers also hope to use it to study the TRAPPIST-1 planets and other planets around M-type stars, too, to figure out whether those planets keep their atmospheres. But even if they maintain their atmospheres, planets around M-type stars won't be much like Earth.

If JWST starts to find that planets around M-type stars don't have atmospheres, then it severely limits what planets in our universe could be habitable. It's not direct observation, but this *could* also tell us about the magnetic fields of these planets. Pair this with observations of protoplanets and other young systems and we might be able to get a few of the puzzle pieces into what makes up a planet.

EXPLORING OUR SOLAR SYSTEM

There are a good handful of upcoming missions that could tell us a lot about our own solar system. In 2023, the Psyche mission will go to the asteroid belt. This isn't the first dedicated mission to it, but its destination is one of its most intriguing worlds. Everything we've learned about the asteroid 16 Psyche indicates that it has a lot of nickel and iron, which means it could be the core of a small planetesimal. If this is the case, it will be one of our few opportunities to explore the actual core of a planet-like object. If it's not a remnant core, then it's an example of an iron-rich planetesimal that may have exciting phenomena, including iron volcanism in the early asteroid belt.[11]

The Mars Sample Return mission, a NASA-ESA collaboration that is still in development, could answer a lot of questions by bringing samples back from the red planet. This could both confirm ancient life by subjecting the samples to thorough experiments in Earth labs and provide Martian materials that tell us about its geologic history, the collapse of its magnetic field, and so much more. Recent *Perseverance* rover data suggest that its landing site, Jezero Crater, experienced an ancient volcanic event, so we could even get materials that hint at what once happened deep below the surface of Mars.

Getting these samples back to Earth will be an entirely different engineering endeavor. NASA will work with the ESA on a complex suite of missions under the Mars Sample Return umbrella. The Sample Retrieval Lander will send out two helicopters—bigger, more powerful versions of *Ingenuity*—to grab samples captured by *Perseverance*. The Sample Transfer Arm, built by ESA, will carefully place the retrieved samples into the Mars Ascent Vehicle, which will send the samples to

Mars orbit.[12] ESA's Earth Return Orbiter will then grab those samples and come back to Earth with the precious cargo in tow.[13]

It sounds like a lot. It *is* a lot. And not all the technology is developed yet. The whole mission framework won't launch until 2027 at the earliest. And when it does launch, getting samples back to Earth will take another five or six years. Only then will we have the first pristine samples from the surface of Mars here on Earth.

The extremely long schedules that missions require—sometimes decades between proposing a mission and its arrival at a planet—mean that several generations of scientists have to build on previous work to see the mission through—sort of a collaboration through time. This is why it's so crucial to maintain a strong training and mentorship program for early career planetary scientists so they can take over the helm of a mission at the appropriate time.[14]

The Mars Sample Return is expected to take up a lot of NASA's resources, but there are still some missions on the docket to help us understand the rest of the solar system, including the interiors of planets and moons.

Beyond *Juno*'s ongoing indirect observations of Jupiter's interior, astronomers will get a few chances to observe other parts of Jupiter's vicinity. The Lucy mission is already en route to the Trojan asteroids of Jupiter. These are groups of asteroids that follow in front of and behind Jupiter without orbiting it, instead migrating to a stable area that keeps them moving along with Jupiter. They represent some of the most pristine materials in the solar system, and even though we aren't sending back an OSIRIS-REx–like sample, we'll be able to gain knowledge of a few things. A thermal spectrometer will tell us about the surface properties of these small building blocks, and infrared instruments will look for ices, silicates, and other chemicals that give a sense of the elemental pantry of the early solar system.

But there are also two missions, one from ESA and one from NASA, that will look at Jupiter's moons and hopefully tell us what's happening deep inside them. While Europa's oceans have pretty direct evidence behind them, the evidence for Ganymede's is a little more indirect. ESA's *Jupiter Icy Moons Explorer*, or *JUICE*, will study Ganymede. The craft launched in April 2023 and will use Earth and Venus gravity assists to arrive at Jupiter in 2031. *JUICE* will first study Jupiter's magnetosphere, then have several flybys of Ganymede, Europa, and Callisto before settling into a Ganymede orbit in 2034.

Some instruments on *JUICE* will study Ganymede's auroras, which will help us understand what's going on deep below and what powers Ganymede's magnetic field. Radar systems will allow the craft to gather data about the subsurface without landing and learn more about it. Gravitational sensors will map the interior of the solar system's largest moon. We won't be able to dig into Ganymede, but it should give us a pretty good sense of what's happening deep below it.

Then there's the *Europa Clipper*, slated to launch in 2024. To ensure its long-term survival, the *Clipper* won't orbit Europa directly but will instead orbit Jupiter and have several encounters with Europa. It should be able to look for chemical signatures on the surface that indicate interactions with the ocean deep below. Ice-penetrating radar will tell us what's under the icy crust of Europa, leading down to the ocean. Plasma and magnetometer instruments could also tell us more about the interactions between Jupiter and Europa, as well as how the ocean interacts with the interior of the moon.

Both those missions are well on their way to launch and are under various states of construction. Another icy moon explorer could take to the skies of Titan, with a potential launch date of 2027. Called Dragonfly, the mission will mainly be examining the surface of Titan. It's a rotocraft that can actually fly thanks to the substantial atmosphere of Titan. However, some areas on the surface may be interacting with materials from deep below, whether through a meteor impact or a cryovolcanic process. This mission could sample those materials and tell us more about the interior of Titan.

The VERITAS (Venus Emissivity, Radio science, InSAR, Topography, and Spectroscopy) craft is one of three missions to Venus, the others being DAVINCI+ (Deep Atmosphere

Venus Investigation of Noble gases, Chemistry, and Imaging)*
and EnVision. VERITAS scientists want to map Venus's surface
and find evidence of volcanism, which could explain the processes that put Venus in its current hellish state. This could
give us evidence of what's happening in the mantle of Venus.
DAVINCI+ is mostly meant to study the atmosphere, but it
will also image the surface. EnVision will investigate Venus's
interior as well as the surface and atmosphere. Together,
these could give us an idea of what is going on beneath Venus,
where we can't exactly send an InSight-inspired mission to
the surface.

NO 2.0

If you take anything from this book, it's the knowledge that our
planet is unique, that the forces that shape it come as much
from below as from above, and that a unique set of circumstances made our life here possible. It's a world we need to
treasure and tend to—a world we need to learn more from. Our
planetary explorations shouldn't be conquests and new do-
mains to settle in—they should be about knowledge and understanding of the world and worlds around us. Given how
many different factors go into making a planet, there really is
no Earth 2.0.

---

*Yes, acronyms for NASA missions are out of control.

Every place we've found thus far is too remote to get to, even for the best propulsion technology. Our technology for long-term survival in space is too primitive, even while Jeff Bezos talks of building colonies in cylindrical ships through his company Blue Origins. All these initiatives remain in the domain of science fiction; any step taken thus far is a baby step, all while temperatures on our own world increase to a level where Earth may be inhospitable in a few decades.

And even if there is a planet with the perfect Goldilocks conditions on it, and we can somehow get there, what right do we have to it? Conquest has changed our planet for the worse, laid waste to whole ecosystems, and decimated entire groups of people. The blueprint for these future endeavors may not be quite as much what we see in *Star Trek* and might be more in line with *The Word for World Is Forest* by Ursula le Guin or *The Sparrow* by Mary Doria Russell, where our own contact and conquest goes horribly awry—for ourselves but also for the other planets. Our own life on Earth hangs in the balance.

There are human ambitions to go to Mars, but some may be folly. Andy Weir's book *The Martian*—and its movie adaptation—deal with a lone astronaut who forges a life on Mars while waiting to be rescued after a botched Mars mission. Thankfully, he has all the resources he needs, but not without ingenious thinking and a whole lot of luck. I should also point out that a soil ecologist I know has major issues with the feasibility of the whole potato thing.

Are We Alone in the Universe?

The question of where—and how—life exists beyond Earth has moti-
vated a great deal of thinking in the twentieth and twenty-first
centuries (among scientists, writers, and average citizens alike), and
it lies at the heart of numerous space-related endeavors including
Mars rovers, missions to icy worlds, and radio sweeps for distant
alien transmissions. Finding traces of life on Mars or on Jupiter's
moon Europa, in whatever forms it might present, could tell us
whether biology is common or rare. But our eight planets, hundreds
of moons, dwarf planets, and millions of other objects orbiting our
Sun can only tell us so much, as all these worlds share variations
of a common chemical lineage. We need to venture into the next
frontiers for a broader research sampling.

The Milky Way comprises 100–400 billion stars, most of which likely
have at least one planet. Many of those exoplanets are very different
from those in our own solar system—miniature versions of Neptune,
worlds that are rocky like Earth but far bigger, unimaginably hot,
giant worlds that orbit their stars in a matter of hours or days. Some
exoplanets may even be the leftover cores of gas giants, called
Chthonian planets. Not all planets are suitable for life—but some
might be, and understanding their conditions is integral to finding
out which are viable candidates. "I really think that the origin of life on
Earth and the search for life elsewhere are two sides of the same
coin," says Sarah Hörst, a Johns Hopkins University professor who
studies planetary atmospheres.

As demonstrated throughout this book, what happens inside a planet is as important as what happens on its surface, if not more so. We have slowly cracked some of the secrets of our own solar system, but the next big goal—one that likely won't even be fully realized in any of our lifetimes—is to figure out what goes on inside exoplanets.

It won't be easy. Right now, we can't really take a good image of an exoplanet; the handful of direct images we have are blobs of light. That makes the prospect of studying the interior seem even *more* difficult—but not impossible.

Hörst mentions two ways that we can find interior processes with current or near-future technology. The first is something NASA, ESA, and the Canadian Space Agency's JWST observatory can do. It has an instrument called NIRSpec (short for "Near Infrared Spectrograph") that can observe exoplanets as they pass in front of their stars, and look for fingerprints of their atmospheres' contents filtering through that intense light. Certain chemicals in those atmospheres—nitrogen, oxygen, and water vapor, for instance—could increase the likelihood of Earth-like conditions.

One chemical in particular could be tied to volcanoes: sulfur dioxide. The JWST telescope has already detected it on a Saturn-sized world called WASP-39b, proving that it's capable of finding it. If we find it on a smaller, rocky world, it could point to volcanoes, which are important, Hörst says, because rocky planets in our solar system relied on volcanism to produce their atmospheres.

Gas giants can hold onto their primordial atmospheres; rocky planets can't, "but they're big enough to do things like have their interior differentiate and have the possibility of plate tectonics, volcanism, and deep-sea vents," Hörst says. Those processes, in turn, are key to life on Earth, especially because the two leading theories for its origin are in deep-sea vents or in warm little ponds on young volcanic islands (for more on how critical volcanoes are to life on Earth, see page 89). Also, if life's origins are at deep-sea thermal vents, an abundance of creatures near them, such as volcanic tube worms or scaly-foot snails, suggests that we need a rethink on the definition of habitable.

But Earth also has a magnetic field that shields us from some of the most dangerous radiation from the Sun. A weak magnetic field on Mars could have doomed its chances for life early on by pulling in the worst the Sun can deliver and wiping out its atmosphere, what Hörst calls the "worst-case scenario for atmospheric loss." The lack of a magnetic field on Venus, incidentally, may not have caused its atmosphere to go away, but it could have led to its runaway greenhouse effect, driving out water. The magnetic fields protect life on Earth, but we need to understand other worlds to know exactly what's the right strength to hold on to Earth-like conditions.

While astronomers are a long way off from being able to directly detect magnetic fields on exoplanets, they *could* be able to glean hints of them with future instruments by seeing how a flare from a star affects its atmosphere.

"At least in our solar system, there are certain atmospheric [chemical] changes that are characteristic of auroral processes because they're higher energy than what's normally happening day to day in that planet's atmosphere," Hörst says. This could, for instance, alter the amount of oxygen or nitrogen seen after a flare hits.

Such studies could get us part of the way to understanding the inner workings of exoplanets and add more information to the puzzles surrounding how life on Earth arose, and whether it happens elsewhere. Finding life—whether as small as microorganisms or something more—could help us understand our place in an unimaginably vast cosmos, generating all-new ways of thinking across a diversity of fields, including science, religion, and philosophy.

FURTHER READING

Gonstral, Aaron. "Exposed in a Warm Little Pond." *Astrobiology at NASA: Life in the Universe.* December 6, 2018. https://astrobiology.nasa.gov/news/exposed-in-a-warm -little-pond/.

Kaltenegger, L., W. G. Henning, and D. D. Sasselov. "Detecting Volcanism on Extrasolar Planets." *Astronomical Journal* 140, no. 5 (2010): 1370. https://doi.org/10.1088 /0004-6256/140/5/1370.

NASA. "NASA Climate Modeling Suggests Venus May Have Been Habitable." News release, August 11, 2016. https://www.nasa.gov/feature/goddard/2016/nasa-climate -modeling-suggests-venus-may-have-been-habitable.

Sakata, R., et al. "Effects of an Intrinsic Magnetic Field on Ion Loss from Ancient Mars Based on Multispecies MHD Simulations." *JGR Space Physics* 125, no. 2 (2020). https://doi.org/10.1029/2019JA026945.

Vidotto, A. A., N. Feeney, and J. H. Groh. "Can We Detect Aurora in Exoplanets Orbiting M Dwarfs?" *Monthly Notices of the Royal Astronomical Society* 488, no. 1 (2019): 633–44. https://doi.org/10.1093/mnras/stz1696.

The actual soil of Mars is laden with toxic chemicals called perchlorates. These chemicals can be found on Earth in some high desert terrain, and their health effects are. . . . not great. They're listed as one of the chief pollutants at the Rockets, Fireworks, and Flares Superfund Site in Rialto, California, and can inhibit lung and thyroid function. These toxic qualities could also affect the ability of plants to grow in the soil, and unless huge remediation efforts somehow become possible, for any sustained human presence on the Red Planet.

We know the perchlorates are present there because of seasonal flows seen on Mars called recurring slope lineae. When Mars gets warm enough, water ice unfreezes. Normally, surface water on Mars would sublimate away in the thin atmosphere. But by combining with the perchlorate salts, the water can flow on the surface, carving certain patterns in areas like the Newton Crater.

Most water on Mars is trapped in the form of ice below the surface. There's no liquid water to be found above. But this water is probably polluted by these perchlorates in some ways, making it non-potable for humans.

Then there's the radiation. Without a functional magnetic field, a Mars settler would be battered with all forms of cosmic radiation, with no modern technology to really remediate it in the long term. While living underground would help, this would hinder the ability to travel to Mars more widely and efficiently.

And Mars is our "best" chance at a life off-world. Humans *might* survive in upper cloud layers of Venus, but it would be hard to sustain that existence. The Moon's south pole, the target of the Artemis program, does have promising assets like water and deep caves to withstand space radiation. Plus, it's close to home. But it's the training wheels for being able to go anywhere.

Beyond the asteroid belt, conditions are both too cold and often too deadly. Europa, for instance, is continuously bathed in the radiation of Jupiter, radiation amounts that require significant shielding just for a robot to make it through. Titan may seem intriguing, but it's also −290 degrees Fahrenheit.

So not only is there no Earth 2.0, but there's really nowhere else for us to go.

But maybe we don't *need* to go anywhere, at least with our own boots. If we limit our goals to the pursuit of scientific knowledge to educate us and benefit life here, however, we can learn more about our origins. We can find the processes that make the various strange and delightful worlds of the outer solar system tick. We can better understand what makes us special, and what unique set of circumstances led us to where we are.

I hope I've been able to portray how wondrous the inner worlds of planets are, including our own home sweet home. The forces at play in a planet's interior create entirely new environments that, although perhaps not as fanciful as the dinosaurs in Jules Verne's classic *Journey to the Center of the Earth*, are still completely foreign to our experiences living on the surface. I hope we continue our fascination with what lies beneath our feet. Each time I see a child on the beach with a shovel, I'll encourage them to keep digging. Who knows what they might find?

Acknowledgments

MANY WONDERFUL MENTORS throughout my education and career have guided and inspired me. Dick van Raadshooven, Jerry Mitrovica, John Percy, Jeremy Bloxham and Maria Zuber: thank you for giving me your time, investing in my development, and sharing your wisdom.

To members of my research group throughout the years: you make the time, effort, and stress all worth it. A special thank-you to my first set of PhD students—Girija Dharmaraj, Ryan Vilim, and Bob Tian—who dealt with my learning the ropes of being a professor at the University of Toronto. Another shout-out to recent group members of MagPI at Johns Hopkins University: Miché Aaron, Regupathi Angappan, Ankit Barik, Cauê Borlina, Viranga Perera, Junellie González Quiles, Mayuri Sadhasivan, Melissa Sims, and Chi Yan—thank you for how much you've taught me and for the privilege of seeing you grow through your journeys as amazing human beings and kick-ass scientists.

To my colleagues in the Department of Earth and Planetary Sciences at Johns Hopkins University: thank you for inviting me to join this wonderful place and working with me to create a vision for what an inclusive, fun, and excellent department should be. A special note of gratitude to Sarah Hörst, Emmy Smith, Maya Gomes, and Meghan Avolio. You know why.

To my ELATES cohort: What an amazing experience it has been this past year to learn with all of you. Thank you to Sharon Walker for leading the program, and thank you to all the amazing mentors and coaches (especially Adrienne Minerick and Jackie Radford) who've shared their wisdom and have shown me where my path is leading. Also a huge thanks to the most amazing learning community I could possibly have asked for: Sarah Bergbreiter, Anna Erickson, Jingjing Li, and Scarlett Miller. You know you are now my family.

To the staff and leadership of the American Geophysical Union: I've been so fortunate to work with you over the years to contribute to such an important mission. Thank you for taking in a shy young professor and showing me the impact and value that working to support our community can have. Special thanks to Cheryl Enderlein, Randy Fiser, Brooks Hanson, Janice Lachance, Christine McEntee, Lauren Parr, and Billy Williams. The insights I've gained from working with all of you have improved my life immensely.

This book would not have been possible without the amazing team at Johns Hopkins University working with the Bloomberg Distinguished Professors. To Denis Wirtz and Julie Messersmith in the Office of Research: you are amazing colleagues and have provided so much support. To John Wenz, without whom I would have given up many times throughout this process: thank you for all your help and patience, and for sharing your wisdom and writing talents on this project. To the Wavelengths program director, Anna Marlis Burgard: thank you for your vision and leadership on this project, and for sharing my love of the intersections of science and art. Thank you also to Nicole Kit for translating my thoughts into the visuals presented in this book. And much appreciation to the rest of the JHU Press team, including senior science acquisitions editor Tiffany Gasbarrini, copyeditor Charles Dibble, and executive director Barbara Kline Pope, who laid the foundation for this book series.

Major thanks to Scott King for providing helpful feedback on this manuscript. I'm glad we can get along even though we have massively different opinions on Venus and chocolate with bacon.

To the Broad Squad—Charlene Gareffa, Brandee Pidgeon, Alwynn Pinard, and Heidi Pitfield—I can't put into words what you all mean to me and how lucky I was to meet you at Marymount those many years ago. Your friendship shaped me into the person I am.

To my family in Canada, thank you for everything. Dad, Maike, Maria, Sara, Kelsie: you mean the world to me. And a special thank-you to my mom, who isn't here with us to read this book, but who I know would have loved it and been very proud of me. I'm always thinking of you.

And finally, to the most amazing partner and best friend I could ever imagine: Tony, thank you for stealing horses with me (an inside joke—no criminal activity or harm has occurred). You mean all these worlds to me.

Notes

CHAPTER 1. GAZING INWARD

1. Sabine Begall et al., "Magnetic Alignment in Grazing and Resting Cattle and Deer," *Proceedings of the National Academy of Sciences* 105, no. 36 (2008): 13451–55, https://doi.org/10.1073/pnas.0803650105.

2. S. F. Odenwald and J. L. Green, "Bracing for a Solar Superstorm," *Scientific American* 299, no. 2 (August 2008): 80–87, https://doi.org/10.1038/scientificamerican0808-80.

3. D.T. Phillips, "Near Miss: The Solar Superstorm of July 2012," *NASA Science*, July 23, 2014, https://science.nasa.gov/science-news/science-at-nasa/2014/23jul_superstorm.

4. Mark Piesing, "During the Cold War, the US and Soviets both created ambitious projects to drill deeper than ever before." *BBC Future*, May 6, 2019, https://www.bbc.com/future/article/20190503-the-deepest-hole-we-have-ever-dug.

5. John Steinbeck, "High drama of bold thrust through the ocean floor: Earth's second layer is tapped in prelude to MOHOLE," *Life Magazine*, April 14, 1961.

6. D. S. Greenberg, "Mohole: The Project That Went Awry," *Science* 143, no. 3602 (1964): 115–19, https://doi.org/ 10.1126/science.143.3604.334.

7. J. Paul, "Cratons, Why Are You Still Here?," *Eos*, March 25, 2021, https://doi.org/10.1029/2021EO156381.

CHAPTER 2. GAZING OUTWARD

1. P. Banerjee, Y.-Z. Qian, A. Heger, and W.C. Haxton, "Evidence from Stable Isotopes and ^{10}Be for Solar System Formation Triggered by a Low-Mass

Supernova," *Nature Communications* 7, (2016): article 13639, https://doi.org/10.1038/ncomms13639.

2. NASA Solar System Exploration, "'Oumuamua," December 19, 2019, https://solarsystem.nasa.gov/asteroids-comets-and-meteors/comets/oumuamua/in-depth/.

3. Denton S. Ebel and Sarah T. Stewart, "The Elusive Origin of Mercury," in *Mercury: The View after MESSENGER*, ed. Sean C. Solomon et al. (Cambridge: Cambridge University Press, 2018), 497–515.

4. William K. Hartmann and Donald R. Davis, "Satellite-Sized Planetesimals and Lunar Origin," *Icarus* 24 (1975): 504–15, https://doi.org/10.1016/0019-1035(75)90070-6.

5. Alan P. Boss, "Formation of Gas and Ice Giant Planets," *Earth Planetary Science Letters* 202, no. 3–4 (2002): 513–23, https://doi.org/10.1016/S0012-821X(02)00808-7.

6. K. Tsiganis et al. "Origin of the Orbital Architecture of the Giant Planets of the Solar System," *Nature* 435 (2005): 459–61, https://doi.org/10.1038/nature03539.

7. D. Nesvorny, "Young Solar System's Fifth Giant Planet?," *Astrophysical Journal Letters* 742, no. 2 (2011): L22, https://doi.org/10.1088/2041-8205/742/2/L22.

8. Konstantin Batygin and Michael E. Brown, "Evidence for a Distant Giant Planet in the Solar System," *Astronomical Journal* 151, no. 2 (2016): 22, https://doi.org/10.3847/0004-6256/151/2/22.

9. Captain W. S. Jacobs, "On Certain Anomalies Presented by the Binary Star 70 Ophiuchi," *Monthly Notices of the Royal Astronomical Society* 15, no. 9 (1855): 228–30, https://doi.org/10.1093/mnras/15.9.228.

10. Peter van de Kamp, "Astrometric Study of Barnard's Star," *Astronomical Journal* 68 (1963): 295–96, https://doi.org/10.1086/108973.

11. Jacob Berkowitz, "Lost World: How Canada Missed Its Moment of Glory," *Globe and Mail*, September 25, 2009, https://www.theglobeandmail.com/technology/science/lost-world-how-canada-missed-its-moment-of-glory/article4290133/.

12. Bruce Campbell et al., "A Search for Substellar Companions to Solar-type Stars," *Astrophysical Journal* 331 (1992): 902–21, https://doi.org/10.1086/166608.

13. Artie P. Hatzes et al., "A Planetary Companion to γ Cephei A," *Astrophysical Journal* 599 (2003): 1383, https://doi.org/10.48550/arXiv.astro-ph/0305110.

14. Michel Mayor and Didier Queloz, "A Jupiter-Mass Companion to a Solar-Type Star," *Nature* 378 (1995): 355–59, https://doi.org/10.1038/378355a0.

CHAPTER 3. TELLTALE PLANETARY PARCELS

1. Office of the Press Secretary at the White House, "President Clinton Statement Regarding Mars Meteorite Discovery," news release, August 7, 1996, https://www2.jpl.nasa.gov/snc/clinton.html.

2. International Polar Foundation, "Antarctic Scientists Discover 18kg Meteorite," news release, February 28, 2013, http://www.antarcticstation.org/news_press/press_release/antarctic_scientists_discover_18kg_meteorite.

3. L.J. Spence, "Hoba (South-West Africa), the Largest Known Meteorite," *Mineralogical Magazine and Journal of the Mineralogical Society* 23, no. 136 (1932): 1–18, https://doi.org/doi:10.1180/minmag.1932.023.136.03.

4. K. G. Gardner-Vandy et al., "Making GRA 06128/9: Chemical Success and Physical Challenges," oral presentation at the Workshop on Planetesimal Formation and Differentiation, Department of Terrestrial Magnetism, Carnegie Institution for Science, Washington, DC, October 27–29, 2013.

5. Samuel H. C. Cabot and Gregory Laughlin, "Lunar Exploration as a Probe of Ancient Venus," *Planetary Science Journal* 1, no. 3 (2020): 66, https://doi.org/10.3847/PSJ/abbc18.

6. V. E. Hamilton et al., "Meteoritic Evidence for a Ceres-Sized Water-Rich Carbonaceous Chondrite Parent Asteroid," *Nature Astronomy* 5 (2021): 350–55, https://doi.org/10.1038/s41550-020-01274-z.

7. Alex Halliday and John Chambers, "The Origin of the Solar System," in *Encyclopedia of the Solar System*, ed. Tilman Spohn, Doris Breuer, and T. V. Johnson (Amsterdam: Elsevier, 2014), 29–54.

8. Hope A. Ishii et al., "Multiple Generations of Grain Aggregation in Different Environments Preceded Solar System Body Formation," *Proceedings of the National Academy of Sciences* 115, no. 26 (2018): 6608–13, https://doi.org/10.1073/pnas.1720167115.

9. Jan D. Kramers et al., "The Chemistry of the Extraterrestrial Carbonaceous Stone 'Hypatia': A Perspective on Dust Heterogeneity in Interstellar Space," *Icarus* 382 (2022): 115043, https://doi.org/10.1016/j.icarus.2022.115043.

10. Steven B. Shirey and Stephen H. Richardson, "Start of the Wilson Cycle at 3 Ga Shown by Diamonds from Subcontinental Mantle," *Science* 333, no. 6041 (2011): 434–36, https://doi.org/10.1126/science.1206275.

11. Tingting Gu et al., "Hydrous Peridotitic Fragments of Earth's Mantle 660 Km Discontinuity Sampled by a Diamond," *Nature Geoscience* 15 (2022): 950–54, https://doi.org/10.1038/s41561-022-01024-y.

12. Byeongkwan Ko et al., "Water-Induced Diamond Formation at Earth's Core-Mantle Boundary," *Geophysical Research Letters* 49, no. 16 (2022), https://doi.org/10.1029/2022GL098271.

13. Rajdeep Dasgupta et al., "Carbon Solution and Partitioning between Metallic and Silicate Melts in a Shallow Magma Ocean: Implications for the Origin and Distribution of Terrestrial Carbon," *Geochimica et Cosmochimica Acta* 102 (2013): 191–212, https://doi.org/10.1016/j.gca.2012.10.011.

CHAPTER 4. FIERCE AND FORMATIVE FORCES

1. Gravity Recovery and Climate Experiment/University of Texas at Austin, "GRACE Gravity Model 05," released 2016, https://www2.csr.utexas.edu/grace/gravity.

2. NASA, "GRAIL's Gravity Map of the Moon," news release, October 4, 2017, https://moon.nasa.gov/resources/55/grails-gravity-map-of-the-moon/.

3. Richard A. Kerr, "Past Tectonics on Mars?," *Science*, April 29, 1999, https://doi.org/10.1126/article.38516.

4. Kerr, "Past Tectonics on Mars?"

5. NASA, "Scientists Find That Saturn's Rotation Period Is a Puzzle," news release, June 28, 2004, https://www.nasa.gov/mission_pages/cassini/media/cassini-062804.html.

6. M. N. Chowdhury et al., "Saturn's Weather-Driven Aurorae Modulate Oscillations in the Magnetic Field and Radio Emissions," *Geophysical Research Letters* 49, no. 3 (2022), https://doi.org/10.1029/2021GL096492.

CHAPTER 5. HOW WE PEER INSIDE PLANETS

1. A. Mittelholz et al., "Timing of the Martian Dynamo: New Constraints for a Core Field 4.5 and 3.7 Ga Ago," *Science Advances* 6, no. 18 (2020), https://doi.org/ 10.1126/sciadv.aba0513.

2. United States Geologic Survey, Earthquake Network Team, "GSN—Global Seismographic Network," accessed February 28, 2023, https://www.usgs.gov/programs/earthquake-hazards/gsn-global-seismographic-network.

3. One could make an argument that Europa, the ice-covered moon of Jupiter, has a form of plate tectonics happening in its ice layer. Simon A. Kattenhorn and Louise M. Prockter, "Evidence for Subduction in the Ice Shell of Europa," *Nature Geoscience* 7 (2014): 762–67, https://doi.org/10.1038/ngeo2245.

4. Alan Shepard and Deke Slayton, *Moon Shot: The Inside Story of America's Race to the Moon,* (Atlanta, GA: Turner, 1994).

5. "Quakes on the Moon," *Berkeley Seismology Lab's Seismo Blog,* published online July 20, 2009, https://seismo.berkeley.edu/blog/2009/07/20/quakes-on-the-moon.html.

6. Renee C. Weber et al., "Seismic Detection of the Lunar Core," *Science* 331, no. 6015 (2011): 309–12, https://doi.org/10.1126/science.1199375.

7. M.P. Panning et al., "Farside Seismic Suite (FSS): Surviving the Lunar Night and Delivering the First Seismic Data from the Farside of the Moon," 53rd Lunar and Planetary Science Conference, Held March 7–11, 2022, at the Woodlands, Texas (2022), https://www.hou.usra.edu/meetings/lpsc2022/pdf/1576.pdf.

8. Ravit Helled and David Stevenson, "The Fuzziness of Giant Planets' Cores," *Astrophysical Journal Letters* 840, no. 1 (2017), https://doi.org/10.3847/2041-8213/aa6d08.

9. NASA Solar System Exploration, "Scientists Finally Know What Time It Is on Saturn," news release, January 18, 2019. https://solarsystem.nasa.gov/news/814/scientists-finally-know-what-time-it-is-on-saturn/; Christopher Mankovich et al., "*Cassini* Ring Seismology as a Probe of Saturn's Interior, I: Rigid Rotation," *Astrophysical Journal* 871, no. 1 (2019), https://doi.org/10.3847/1538-4357/aaf798.

10. John D. Anderson and Gerald Schubert, "Saturn's Gravitational Field, Internal Rotation, and Interior Structure," *Science* 317, no. 5843 (2007): 1384, https://doi.org/10.1126/science.1144835.

11. Alex S. Konopliv et al., "Detection of the Chandler Wobble of Mars from Orbiting Spacecraft," *Geophysical Research Letters* 47, no. 21 (2020): e2020GL090568, https://doi.org/10.1029/2020GL090568.

12. Amy C. Barr and Robin M. Canup, "Origin of the Ganymede-Callisto Dichotomy by Impacts during the Late Heavy Bombardment," *Nature Geoscience* 3 (2010): 164–67, https://doi.org/10.1038/ngeo746.

CHAPTER 6. CURIOUS PLANETARY ELEMENTS

1. David Kramer, "Helium Is Again in Short Supply," *Physics Today,* published online April 4, 2022, https://physicstoday.scitation.org/do/10.1063/PT.6.2.20220404a/full/.

2. J. E. Lupton and H. Craig, "A Major Helium-3 Source at 15{degrees}S on the East Pacific Rise," *Science* 214, no. 4516 (1981): 13–18, https://doi.org/10.1126/science.214.4516.13.

3. Zhihua Xiong et al., "Helium and Argon Partitioning between Liquid Iron and Silicate Melt at High Pressure," *Geophysical Research Letters* 48, no. 3 (2021), https://doi.org/10.1029/2020GL090769.

4. Peter L. Olson and Zachary D. Sharp, "Primordial Helium-3 Exchange between Earth's Core and Mantle," *Geochemistry, Geophysics, Geosystems* 23, no. 3 (2022), https://doi.org/10.1029/2021GC009985.

5. S. Brygoo et al., "Evidence of Hydrogen-Helium Immiscibility at Jupiter-Interior Conditions," *Nature* 593 (2021): 517–21, https://doi.org/10.1038/s41586-021-03516-0; Hugh F. Wilson and Burkhard Militzer, "Sequestration of Noble Gases in Giant Planet Interiors," *Physical Review Letters* 104, no. 12 (2010): 121101, https://doi.org/10.1103/PhysRevLett.104.121101.

6. Ronald Redmer et al., "The Phase Diagram of Water and the Magnetic Fields of Uranus and Neptune," *Icarus* 211, no. 1 (2011): 798–803, https://doi.org/10.1016/j.icarus.2010.08.008.

7. Marius Millot et al., "Nanosecond X-ray Diffraction of Shock-Compressed Superionic Water Ice," *Nature* 569 (2019): 251–55, https://doi.org/10.1038/s41586-019-1114-6.

8. Christophe Sotin et al., "Titan's Interior Structure and Dynamics after the Cassini-Huygens Mission," *Annual Review of Earth and Planetary Sciences* 49 (2021): 579–607, https://doi.org/10.1146/annurev-earth-072920-052847.

9. James Stevenson, Jonathan Lunine, and Paulette Clancy, "Membrane Alternatives in Worlds without Oxygen: Creation of an Azotosome," *Science Advances* 1, no. 1 (2015), http://doi.org/DOI: 10.1126/sciadv.1400067.

10. Gaël Choblet et al., "Powering Prolonged Hydrothermal Activity inside Enceladus," *Nature Astronomy* 1 (2017): 841–47, https://doi.org/10.1038/s41550-017-0289-8.

11. Alyssa Rose Rhoden and Matthew E. Walker, "The Case for an Ocean-Bearing Mimas from Tidal Heating Analysis," *Icarus* 376 (2022): 114872, https://doi.org/10.1016/j.icarus.2021.114872.

12. J. H. Eggert et al., "Melting Temperature of Diamond at Ultrahigh Pressure," *Nature Physics* 6, (2010): 40–43, https://doi.org/10.1038/nphys1438.

13. Zhiyu He et al., "Diamond Formation Kinetics in Shock-Compressed C–H–O Samples Recorded by Small-Angle X-Ray Scattering and X-Ray Diffraction," *Science Advances* 8, no. 35 (2022), https://doi.org/10.1126/sciadv.abo0617.

14. Eggert et al., "Melting Temperature of Diamond."

15. P. E. Nissen, "The Carbon-to-Oxygen Ratio in Stars with Planets," *Astronomy & Astrophysics* 552 (2013): A73, https://doi.org/10.1051/0004-6361/201321234.

16. E. Stonkutė et al., "High-resolution Spectroscopic Study of Dwarf Stars in the Northern Sky: Lithium, Carbon, and Oxygen Abundances," *Astronomical Journal* 159, no. 3 (2020), https://doi.org/10.3847/1538-3881/ab6a19.

17. Marc J. Kuchner and Sara Seager, "Extrasolar Carbon Planets," *Arxiv.org* astrophysics preprint, 2005, https://doi.org/10.48550/arXiv.astro-ph/0504214.

18. NASA Exoplanet Exploration, A Tale of Two Worlds: Silicate versus Carbon Planets (Artist Concept)," news release, December 15, 2022, "https:// exoplanets.nasa.gov/resources/177/a-tale-of-two-worlds-silicate-versus -carbon-planets-artist-concept/.

19. Nikku Madhusudhan et al., "A Possible Carbon-Rich Interior in Super-Earth 55 Cancri E," *Astrophysical Journal Letters* 759, no. 2 (2012): L40, https://doi .org/10.1088/2041-8205/759/2/L40.

20. Robin George Andrews, "NASA to Seek Iron-Spewing Volcanoes at Psyche," *Scientific American,* posted online April 25, 2019, https://www.scientificamerican .com/article/nasa-to-seek-iron-spewing-volcanoes-at-psyche/.

CHAPTER 7. THE FUTURE OF PLANETARY EXPLORATION

1. E. M. Standish, "Pluto and Planets X," *Completing the Inventory of the Solar System: Astronomical Society of the Pacific Conference Proceedings* 107, ed. T. W. Rettig and J. M. Hahn (1996): 163–70.

2. Greg Buchwald et al., "Pluto Is Discovered Back in Time," *Amateur-Professional Partnerships in Astronomy: Astronomical Society of the Pacific Conference Proceedings* 220, ed. John R. Percy and Joseph B. Wilson (2000): 355.

3. J. B. Holberg, "The Discovery of the Existence of White Dwarf Stars: 1862 to 1930," *Journal for the History of Astronomy* 40, no. 2 (2009): 137–54, https://doi .org/10.1177/002182860904000201.

4. Eric Betz, "What Will Happen to the Planets When the Sun Becomes a Red Giant?," *Astronomy,* September 2020.

5. NASA, "Dead Star Caught Ripping Up Planetary System," news release. June 15, 2022, https://www.nasa.gov/feature/goddard/2022/hubble-dead-star -caught-ripping-up-planetary-system.

6. NASA, "Dead Star Caught Ripping Up Planetary System."

7. Samuel K. Grunblatt et al., "TESS Giants Transiting Giants, II: The Hottest Jupiters Orbiting Evolved Stars," *Astronomical Journal* 163, no. 3 (2022): 120, https://doi.org/10.3847/1538-3881/ac4972.

8. David Ehrenreich et al., "A Giant Comet-like Cloud of Hydrogen Escaping the Warm Neptune-Mass Exoplanet GJ 436b," *Nature* 522 (2015): 459–61, https://doi.org/10.48550/arXiv.1506.07541.

9. A. Léger et al., "The Extreme Physical Properties of the Corot-7b Super-Earth," *Icarus* 213, no. 1 (2011): 1–11, https://doi.org/10.1016/j.icarus.2011.02.004.

10. Chanda Prescod-Weinstein et al., "The James Webb Space Telescope Needs to Be Renamed," *Scientific American,* March 1, 2021, https://www.scientificamerican.com/article/nasa-needs-to-rename-the-james-webb-space-telescope.

11. Michael K. Shepard et al., "Asteroid 16 Psyche: Shape, Features, and Global Map," *Planetary Science Journal* 2, no. 4 (2021): 125, https://doi.org/10.3847/PSJ/abfdba.

12. NASA Science Mars Exploration, "NASA Will Inspire World When It Returns Mars Samples to Earth in 2033," news release, July 27, 2022, https://mars.nasa.gov/news/9233/nasa-will-inspire-world-when-it-returns-mars-samples-to-earth-in-2033/

13. NASA Science Mars Exploration, NASA's Angie Jackman Works to Develop Rocket That Will Bring Mars Samples to Earth," news release, March 8, 2022, https://mars.nasa.gov/news/9141/nasas-angie-jackman-works-to-develop-rocket-that-will-bring-mars-samples-to-earth/.

14. B. Fernando et al., "Inclusion of Early-Career Researchers in Space Missions," *Nature Astronomy* 6, no. 12 (2022): 1339–41, https://doi.org/10.1038/s41550-022-01861-2.

Index

accretion, planetary, 41–42

achondrites, 72, 73–74, 75

acoustics, auroral, 7

acrylonitrile, 176

Adams, Walter, 191

Ahuna Mons (Ceres), 188

Aldrin, Buzz, 128

Allan Hills 84001 (meteorite), 66–67

alpha particles, 150–51

altitude and variations in gravity, 98–100

American Miscellaneous Society, 18–19

ammonia: and cryovolcanism on Ariel and Miranda, 188; and giant planets, 39, 51, 97, 167, 168, 171–72; on Pluto, 178

amphiboles, 73

angular momentum, 35–38, 115, 139

Antarctica and meteorites, 66–73, 75

Antarctic Search for Meteorites, 66–67

Apollo missions, 83, 127–28

Argyre (Mars), 101, 121

Ariel, 188

Armstrong, Neil, 128

Arrokoth, 63–65

Artemis program, 215

asteroids: defined, 44; detecting, 82; as leftovers from formation of solar system, 44, 54–55; meteorites from, 70, 72, 75, 77, 82; mining of, 78, 80; missions to, 44, 78–81, 184–85, 202–3, 204–5; as planetesimals, 44, 78–79. *See also* Ceres; Trojan asteroids

asthenosphere, 15, 100, 113

atmosphere (Earth): atmospheric pressure, 21–22; and carbon dioxide, 179–80; magnetic field as protecting, 31; and plate tectonics, 113

atmospheres on other planets and moons: and carbon dioxide, 180; on exoplanets, 60–61, 196, 200–201, 210–13; and life, 210–13; Mars, 180, 201, 211; Mercury, 32; Saturn, 160–61; stripping of, 196, 200–201, 211; Titan, 31–32; Venus, 31–32, 146–47, 180, 207, 211

atmospheric pressure, 21–22

aurora borealis, 1–2, 6–7, 9–13

axisymmetry, 117, 161–62

Barnard's Star, 56

boreholes, 19, 23

Borisov, 84–85

buoyancy: and convection, 4–5, 125, 151; and crust, 26; and differentiation, 47

Callisto, 55, 103, 144–46, 174, 175, 205

Campbell, Bruce, 56–57

55 Cancri, 183

Cao, Hao, 163–65

carbon: carbon cycle, 2, 89, 111, 113; in core, 27, 88–89; diamonds, 85–87, 180–82, 188; and formation of planets, 39, 51; graphene, 24, 26; and life on Earth, 179–80; liquid hydrocarbons on Titan, 176; in mantle, 88–89; in meteorites, 75, 84; ratio to oxygen in star systems, 182–83

carbon dioxide: in atmospheres, 147, 179–80, 196; and formation of planets, 39; and limnic eruptions, 186

carbon planets, 183

Carrington, Richard, 12

Carrington event, 12

Cassini mission, 114, 118–19, 134, 136, 163–64, 177

Center for Near-Earth Objects Studies, 82

Ceres, 44, 45, 179, 187–88

Charon, 95, 168, 178–79

Chicxulub Crater, x

chondrites, 72, 73–75, 76

chondrules, 72

Chthonian planets, 209

climate change, 180

climate indicators, 99

Clinton, Bill, 67

Cohen, Barbara, 68–72

comets, 44, 53, 54–55, 84

compass, development of, 8

conservation law of angular momentum, 35–38, 115, 139

continental drift, 109. *See also* plate tectonics

convection: and buoyancy, 4–5, 125, 151; Earth's core, 4–5, 27; Earth's mantle, 111–12, 125, 151–53, 155–56; Jupiter, 158, 164–65; Mars, 121, 132; Mercury, 106; Neptune, 169–73; and plate tectonics, 111, 113, 152–53; Saturn, 158, 160, 164–65; stagnant lid convection, 112; Sun, 133; Uranus, 169–73; Venus, 32, 146–47. *See also* dynamos

core (Earth): carbon in, 27, 88–89; composition of, 27–28; convection in, 4–5, 27; in diagram of Earth's layers, 15; differentiation and formation of, 47; dynamo of, 3–4, 9–11, 106, 111; inner, 15, 27–28; outer, 15, 27; radius of, 123; seismic waves in, 123–24

cores of other planets and objects: Callisto, 144; Chthonian planets, 209; differentiation and formation of, 47; diffuse cores, 165; Enceladus, 177; fuzzy cores, 135; Ganymede, 144; Jupiter, 135, 163, 165; Mars, 46, 132, 141; Mercury, 42, 48, 106–8; Moon, 48–50, 128; Saturn, 135, 165; 16 Psyche, 185; terrestrial planets, 47–50; Titan, 32, 176; Venus, 32, 46, 146–47

coronal mass ejections, 11–13

CoRoT-7b (exoplanet), 196

cratons, 86

crust (Earth): age of, 25–26; described, 24–26; in diagram of Earth's layers, 15; and diamonds, 86; explorations of, 17–19, 20; formation of, 25; magnetization of and understanding plate tectonics, 110–11; thickness of, 24–25

crusts of other planets and objects: age of, 25–26; and moonquakes, 128; and terrestrial planets, 42

cryomagma, 188

cryovolcanism, 177, 178, 184, 187–88

cryptochromes, 8

Curiosity (rover), 120

DART (Double Asteroid Redirection Test), 44

DAVINCI+ (Deep Atmosphere Venus Investigation of Noble gases, Chemistry, and Imaging) mission, 206–7

Davis, Donald R., 48

Dawn mission, 187–88

density of interiors, determining, 92–93, 96–97

deserts and meteorites, 69–70

diamond anvil cell, 87–89

diamonds, 85–87, 180–82, 188

differential rotation, 133

differentiation of core, 47

Double Asteroid Redirection Test (DART), 44

Dragonfly mission, 126, 206

dust: defined, 43; and formation of planets, 39–41; and formation of rocks, 43; and formation of solar system, 33–35, 38–41; GEMS in comet dust, 84

dwarf planets, 44, 54–55, 73, 200–201

dynamos: detecting, 106; Earth, 3–4, 9–11, 106, 111; Jupiter, 158; and magnetic fields, 3–4, 9–11, 105–8, 111; Mars, 120–21, 132, 201; Mercury, 106–8; modeling, 170; Neptune, 169–72; physics of, 3; Saturn, 158; Titan, lack of, 3; Uranus, 169–72; Venus, lack of, 3, 146–47

Eagle Nebula, 198

Earth: and carbon cycle, 2, 89, 111, 113; composition of, 42; death of, 193–95; diagram of layers, 15; dynamo of, 3–4, 9–11, 106, 111; explorations of interior, 16–20; explorations of interior, challenges in, 21–24; gravitational acceleration of, 97; gravity, variations in, 97–101, 134; impact craters on, x; as imperfect sphere, 14, 98–100; layers of, 14, 15, 24–28; mass, distribution of, 97–98; mass of, 93; plate tectonics, 87, 108–14, 149–55; precession of, 140–41; radius of, 14; rotational bulge of, 98–100, 115–16; seismic forces on, 122–25; surface pressure, 32. *See also* atmosphere (Earth); core (Earth); crust (Earth); magnetic field of Earth; mantle (Earth)

Earth 2.0, 207–15

earthquakes, *See* seismic forces

Earth Return Orbiter, 204

electromagnetism as formative force, 91, 105–19. *See also* magnetic fields of other planets and objects

electrostatic force, 40–41

elements: and formation of planets, 39, 47, 150; and formation of solar system, 30, 33, 39; in meteorites, 74–75

Enceladus, 45, 168, 177

EnVision mission, 207

epicenter of earthquakes, 124

Epsilon Eridani, 191

Europa: discovery of, 103; formation of, 55; influence on orbit of Io, 104; missions to, 175, 205, 206; as uninhabitable, 215; water on, 45, 168, 174–75, 205, 206

Europa Clipper mission, 175, 206

European Space Agency, 201, 203–4, 210

exoplanets: atmospheres on, 60–61, 196, 200–201, 210–13; carbon planets, 183; death of, 195–96; defined, 45; detecting, 56–62; formation of, 57–59; Hot Jupiters, 46, 60; life on, 60–61, 209–12; magnetic fields, 211–12; numbers of, 60; precovery of, 191–92; size of, 60; super-Earths, 46, 60, 61, 195; volcanism on, 210; water on, 60–61, 202

exosphere, 22

exploration: challenges in, 21–24, 163; of Earth's interior, 16–20; future missions, 197–216. See also *specific missions*

Extremely Large Telescope, 200

Farside Seismic Suite, 129

ferrovolcanism, 79–80, 184–85, 203

55 Cancri, 183

51 Pegasi b, 58, 59

fossils and plate tectonics, 109

freezing: of iron in Earth's core, 27–28. *See also* ice

friction: frictional heating, 160; and ice, 36–38; and tidal warming, 103

fusion crust on meteorites, 70

fuzzy cores, 135

G238-44 (white dwarf star), 195

Galileo Galilei, 103, 134, 190

Galileo mission, xviii, 105, 165, 174

Gamma Cephei Ab, 57

Ganymede: core, 144; discovery of, 103; formation of, 55; influence on orbit of Io, 104; magnetic field, 205; missions to, 205–6; moment of inertia, 144–46; water on, 174, 175, 205

gas giants: Chthonian planets, 209; defined, 45–46; determining density of, 97; formation of, 52; and fuzzy cores, 135; Hot Jupiters, 46, 60; metallic hydrogen in, 156–59; seismic forces on, 132–35. *See also* Jupiter; Saturn

GEMS (glass with embedded metal and sulfides) in comet dust, 84

geysers, 175, 186, 187

giant planets: and ammonia, 39, 51, 97, 167, 168, 171–72; determining rotation period, 114–19, 135–36, 137; formation of, 50–53, 150, 160; formation of exoplanets, 58–59; and helium, 51, 97, 135, 150, 156, 160; and hydrogen, 51, 97; and methane, 51, 97, 167, 168, 171–72, 181; missing fifth giant, 52–53; and rotational bulge, 137; seismic forces on, 132–35; and water, 51, 97, 167–69. *See also* gas giants; ice giants

GJ 436 b (exoplanet), 196

glass with embedded metal and sulfides (GEMS) in comet dust, 84

Gliese 436 (star), 196

Global Seismographic Network, 123–25

Goldilocks zone, 61

GRA 06128 (meteorite), 72

GRA 06129 (meteorite), 72

GRACE mission, 98, 99, 134

GRAIL mission, 101, 134

graphene, 24, 26

gravity: and asteroids, 44; and dwarf planets, 44; on Earth, variations in, 97–101, 134; as force, 91–105; and formation of planets, 41, 51; and formation of solar system, 35–38, 41; gravitational acceleration, 96, 97; gravitational locking, 104; laws of, 94;

maps, 98, 100–101; mass, relation to, 92–95, 100; mass of planets, determining with, 93–97, 146; of Moon, 101, 102–3, 134; and planetary migration, 59; and precession, 140; and rotational bulge, 98–100; and Saturn ring seismology, 134–35; self-gravity, 101; of 16 Psyche, 80; of Sun, 101–3; and tides, 101–5

Greenberg, D. S., 19

greenhouse gases, 179–80

habitable planets, 60–61, 207–15

habitable zone, 61

Halley's Comet, 44

Hartmann, William K., 48

heat: and convection, 4–5, 125, 151; and formation of planets, 47, 50–51, 58; and formation of solar system, 39; and helium rain, 159–60; and mass distribution, 145–46; and planetary differentiation, 47; and planet death, 193–94; and plate tectonics, 113–14; from radioactive decay, 151, 193–94; and tides, 103–5; viscous heating, 160. *See also* temperature

helioseismology, 133

helium: and ^{3}He on Earth, 153–56; creation of through radioactive decay, 150–51, 153, 154; and fuzzy cores, 135; and giant planet formation, 51, 97, 135, 150, 156, 160; helium rain, 159–62; and helium (*cont.*)

plate tectonics, 149–53; primordial, 153–54; and shortage of liquid helium, 149–50; in stars, 46; in Sun, 160–61; in van Maanen's Star, 192

Hellas (Mars), 101, 121

Hess, Harry, 110

Hoba meteorite, 71

Hörst, Sarah, 209–12

Hot Jupiters, 46, 60

Hubble telescope, 38–39, 43, 63–64, 197, 198

hydrogen: and formation of planets, 39, 51, 97; and formation of solar system, 33; and fuzzy cores, 135; and helium rain, 159–62; in ionic water, 169; isotopes of, 83; metallic, 156–59; as most abundant element, 150; in stars, 46

Hypatia (meteorite), 84

ice: and comets, 44; cryovolcanism, 177, 178, 184, 187–88; defined, 51; phases of water and ices on Neptune and Uranus, 167–73; and rocks, 43

ice giants: and ammonia, 39, 51, 97, 167, 168, 171–72; defined, 45; determining density of, 97; and dynamos, 169–72; formation of, 52; and methane, 51, 97, 167, 168, 171–72, 181; phases of water and liquids on, 167–73. *See also* Neptune; Uranus

ice I_c, 168

ice I_h, 167

ice IX, 168

ice XI, 168

impacts and impact craters: Earth, x; Ganymede and Callisto, 145–46; Mars, 101, 121; Moon, 49, 101, 138

inertia, moment of, 142–46

Infrared Astronomical Satellite (IRAS), 191

infrared light, 191, 197–98, 210

InSight mission, 120–21, 130–32, 141–42

International Continental Scientific Drilling Program, 20

International Ocean Discovery Program, 20

interstellar space: boundary with solar system, 16; meteorites from, 84–85. *See also* exoplanets

Io, 55, 103–5, 174, 175, 184

ionic water, 169–71

IRAS (Infrared Astronomical Satellite), 191

iron: in core of Earth, 27–28; in core of Mercury, 42, 81; in cores of terrestrial planets, 47; ferrovolcanism and 16 Psyche, 79–80, 185, 203; in meteorites, 55, 71, 73, 75–77

ironclad sea pangolin snail, 177–78

iron meteorites, 55, 71, 73, 75–77

isotopes, 83–84

Jacob, William Stephen, 56

Jezero Crater (Mars), 203

Johnson, Ted, 195

JUICE (Jupiter Icy Moons Explorer) mission, 205–6

Juno mission, 162, 163, 204

Jupiter: core, 135, 163, 165; dynamo, 158; formation of, 50–52; as gas giant, 52; helium in, 156–59, 160–61, 162; helium rain, 162; magnetic field, 158, 163, 164–65; mass of, 93; metallic hydrogen in, 156–59; missions to, xviii, 105, 162, 163, 165, 174, 204–6; radiation from, 163, 164, 215; size of, 163; winds, 164–65

Jupiter, moons of: discovery of, 103–4; formation of, 55; number of, 45. *See also* Callisto; Europa; Ganymede; Io

Jupiter Icy Moons Explorer (JUICE) mission, 205–6

JWST telescope, 38–39, 43, 197–202, 210

Kepler, Johannes, 94

Kepler-10b, 196

Kepler-16b, 61

Kepler's laws of planetary motion, 94–95, 146

kimberlites, 86, 188

Kobayashi, Issa, 13

Kola Superdeep Borehole, 19

Kontinentales Tiefbohrprogramm der Bundesrepublik Deutschland (KTB), 20

Kuiper Belt, 63–65, 178

Laine, Unto, 7

lava, 151–52, 154

layer cake model of mantle, 155

life on other planets, 60–61, 174, 180, 209–12

limnic eruptions, 186

lithium, 75

lithosphere: in diagram of Earth's layers, 15; and earthquakes, 122; and plate tectonics, 108–14, 152–53; and stagnant lid convection, 112; thickness of, 112; and variations in gravity, 100

Lucy mission, 204–5

Lunar Flashlight mission, 68

lunar tides, 102–3

Luyten, Willem Jacob, 191

Magellan mission, 187

magma, 151–52, 184

magnesium, 39, 75, 76, 182, 195

magnetic field of Earth: animal perception of, 5, 7–8; and aurora borealis, 1–2, 6–7, 9–13; and dynamo of core, 3–4, 9–11, 106, 111; and navigation, 5, 7–8; and plate tectonics, 108–14; as shield, 3, 6, 9–11, 31, 114, 211

magnetic fields of other planets and objects: exoplanets, 211–12; as formative force, 91, 105–19; Ganymede, 205; Jupiter, 158, 163, 164–65; Mars, 120–21, 201; Mercury, 32, 106–8; Neptune, 167, 169; and rotation period, 114–19;

magnetic fields of other planets and objects (*cont.*)

Saturn, 114–19, 158, 161–62, 164–65; 16 Psyche, 80; and stripping of atmosphere, 200–201, 211; Sun, 16, 133; Uranus, 167, 169; Venus, lack of, 211; and water on moons of Jupiter, 175–76

magnetite, 8

magnetometers, 80, 106, 163

magnetoreception, 5, 7–8

magnetosphere (Earth), 6, 9–11

mantle (Earth): carbon in, 88–89; convection in, 111–12, 125, 151–53, 155–56; described, 26–27; in diagram of layers, 15; and diamonds, 86; differentiation and formation of, 47; D″ layer, 125; explorations of, 16–20; layer cake model, 155; lower, 15; and plate tectonics, 111, 113, 152–53; plum pudding model, 156; size of, 26; slab graveyard, 125, 155; as solid, 151–52; upper, 15; water in upper/lower boundary, 87, 88–89

mantles of other planets and objects: differentiation and formation of, 47; Mercury, 42, 48, 50; Moon, 42, 128; and stagnant lid convection, 112; Venus, 42, 207

Marianas Trench, 110

Mariner 10 mission, 107

Mars: atmosphere of, 180, 201, 211; colonization of, 208, 214–15; composition of, 42; core, 46, 132, 141; crust, 25–26; drilling on, 142; and dynamo, 120–21, 132, 201; gravity variations on, 101; impact craters, 101, 121; lithosphere, 112; magnetic field, 120–21, 201; meteorites from, 66–67, 70, 72, 73, 75, 77–82, 84; missions to, xviii, 101, 120–21, 129, 130–32, 141–42, 203–4; plate tectonics on, 112; precession of, 141; rotation of, 121, 141; seismic forces on, 121, 126, 129–32; as terrestrial planet, 42, 47; volcanism on, 187, 203; water on, 214

Mars Global Surveyor, 101

Mars Odyssey mission, 101

Mars Reconnaissance Orbiter mission, 101

Mars Sample Return mission, 203–4

mass: determining planet's mass, 93–97, 146; distribution of, 97–98, 137, 143–44; of Earth, 93, 97–98; and gravitational acceleration, 96; and moment of inertia, 143–44; relation to gravity, 92–95, 100

Mayor, Michael, 58

Mercury: atmosphere, lack of, 32; core, 42, 48, 81, 106–8; crust, 26; determining mass of, 96; dynamo of, 106–8; magnetic field, 32, 106–8; mantle, 42, 48, 50; as terrestrial planet, 42, 47–50; year length, 94

metals: defined, 157; and formation of planets, 39, 50; and GEMS in comet dust, 84; metallic hydrogen, 156–59; metal volcanism and 16 Psyche, 79–80, 184–85, 203; properties of, 157–58

meteorites: from asteroids, 70, 72, 75, 77, 82; composition of, 73–77, 83–84; defined, 44; determining origin of, 77–85; finding, 66–73, 75, 82; with interstellar origins, 84–85; iron meteorites, 55, 71, 73, 75–76; from Mars, 66–67, 70, 72, 73, 75, 77–82, 84; from Moon, 70, 72, 73, 75, 83, 84; on Moon, 72–73; numbers per year, 73; size of, 71; stony irons, 73, 76–77; and understanding planets, 55, 66, 85

meteoroids, 44

meteors, 44, 128

methane: and ice giants, 51, 97, 167, 168, 171–72, 181; and liquid diamond, 181; and Titan, 176

methanol, 188

mid-Atlantic ridge, 110

migration, animal, 5, 7–8

migration, planetary, 59

Mimas, 178

mini-Neptunes, 60

mining, asteroid, 78, 80

Miranda, 188

mirror point, 11

miscibility, 159

Mitrovica, Jerry, xvi–xvii, xviii

Mohole, Project, 17–19

Mohorovičić discontinuity, 17–19

molecular cloud core, 33–35, 38–39

moment of inertia, 142–46

momentum, angular, 35–38, 115, 139

Moon: and collisions, 48–50; core, 48–50, 128; core-mantle boundary, 128; crust, 25–26; formation of, 45, 48–50, 138; gravity of, 101, 102–3, 104; as habitable, 215; impact craters, 49, 101, 138; mantle, 42, 128; meteorites from, 70, 72, 73, 75, 83, 84; meteorites on, 72–73; missions to, 68, 83, 101, 127–29; and precession of Earth, 140; rotational bulge, 138–39; rotational speed and distance from Earth, 138–39; seismic forces on, 126, 127–29; as terrestrial planet, 42, 48–50; and tides, 102–3; volcanism on, 187

moons: defined, 45; and determining mass of planets, 95, 96; as leftovers from formation of solar system, 55; as planetesimals, 55; *vs.* planets, 42; size of, 45. See also *specific moons*

motion: Kepler's laws of planetary motion, 94–95, 146; Newton's laws of, 94, 143

mountains, 100

M-type stars, 202

mud volcanoes, 184, 188

NASA, See *specific missions*

navigation apps, 5

Near Infrared Spectrograph (NIRSpec), 210

neon, 161

Neptune: convection on, 169–73; dynamo, 169–72; formation of, 50–53; liquid diamond in, 181–82; magnetic field, 167, 169; magnetic poles, 162; phases of water and ices on, 167–73; precovery of, 190; proximity to Sun when formed, 52; year length, 94

Neptune, moons of, *See* Triton

New Horizons mission, 63–64

Newton, Isaac, 94

Newton's laws of motion, 94, 143

Nice model, 52

nickel, 27, 78, 203

NIRSpec (Near Infrared Spectrograph), 210

nitrogen: and aurora borealis, 6, 10, 11; exoplanets, 195, 210, 212; and formation of planets, 39, 51; in meteorites, 75; on Triton, 178

noble gases, 75, 161

northern lights, *See* aurora borealis

north star, 140

nutation, 140–41

oceans outside of Earth: Enceladus, 45; Europa, 45; search for, 173–79

OGLE-2005-BLG-390Lb (exoplanet), 61

olivine, 74, 76–77

70 Ophiuchi, 56

orbits: backwards orbit of Triton, 178; elongated orbit of Io, 104; and Kepler's laws of planetary motion, 94–95, 146; and meteorites, 82; and year length, 94

OSIRIS-REx mission, 44

'Oumuamua, 41, 84–85

Outer Space Treaty, 78

oxygen: and aurora borealis, 6, 10, 11; in Earth's core, 27; on exoplanets, 195, 210, 212; and formation of planets, 39, 51; isotopes, 83–84; in Mars' core, 132; in meteorites, 75, 83–84; in phases of water, 167, 169, 173; ratio to carbon in star systems, 182–83; and silicates, 182

pallasites, 76

Parker Solar Probe, 16

Payloads and Research Investigations of the Surface of the Moon (PRISM), 129

Peale, Stan, 104

pebbles and formation of planets, 41

51 Pegasi b, 58, 59

perchlorates, 214

Perkin, Sam, 7

Perseverance (rover), 203

photons, 11

Piccard, Jacques, 110

Planet 9, 53

planetary embryos, 34, 42, 49

planetary interiors: challenges of studying, xix, 21–24, 126, 130; explorations of Earth's interior, 16–20; importance of studying, xix–xx. *See also* core (Earth); cores of other planets and objects;

dynamos; mantle (Earth); mantles of other planets and objects

planetary migration, 59

planetesimals: Arrokoth as, 63; asteroids as, 44, 78–79; and collisions, 41, 49; defined, 44; and formation of solar system, 34, 41–42; and formation of terrestrial planets, 47; meteorites from, 55, 75–76, 77; moons as, 55; 16 Psyche as, 78–79

planets: and accretion, 41–42; death of, 193–97; defined, 45–46; and formation of solar system, 34, 39–42; habitable, 60–61, 207–15; Kepler's laws of planetary motion, 94–95, 146; life on other planets, 60–61, 174, 180, 209–12; *vs.* moons, 42; observing births of, 198–202; protoplanets, 45, 203; types of, 45. *See also* exoplanets; gas giants; giant planets; ice giants; terrestrial planets; *specific planets*

plate tectonics, 87, 108–14, 149–55

plum pudding model of mantle, 156

Pluto, 50, 95, 168, 178–79, 190

Pluto's moon, *See* Charon

Polaris, 140

poles: and determining rotation period, 116–19; Earth, 9, 117, 162; Enceladus, 177; multipolar fields, 169; Neptune, 162, 169; offset of geographical and magnetic poles, 117, 162; Saturn, 117, 161–62; Uranus, 162, 169

precession, 139–42, 144, 146

precovery, 190, 191–92

pressure: and diamonds, 86, 88; Earth's atmospheric, 21–22; in Earth's core, 28; in Earth's mantle, 26–27; Earth's surface pressure, 32; as exploration challenge, 21–24; and metallic hydrogen, 157–58; surface pressure, 31–32

pressure waves, *See* P-waves

primary waves, *See* P-waves

primordial contact binary, Arrokoth as, 63

PRISM (Payloads and Research Investigations of the Surface of the Moon), 129

Project Mohole, 17–19

protoplanetary disks: and formation of exoplanets, 58; and formation of giant planets, 50–52, 150, 160; and formation of solar system, 34, 35–41; and formation of terrestrial planets, 47, 154

protoplanets, 45, 203

proto-Sun, 34, 35, 39

Proxima Centauri, 200

16 Psyche, 78–81, 184–85, 202–3

Psyche Asteroid mission, 78–81, 184–85, 202–3

P-waves, 124

quakes, *See* seismic forces

Queloz, Didier, 58

Quick, Lynnae, 187–88

radial velocity method of exoplanet detection, 57, 58

radiation: and Europa, 215; as exploration challenge, 163; and Jupiter, 163, 164, 215; magnetic field of Earth as shield from, 6, 114, 211; and Mars, 214; and Moon, 215; and Saturn, 164, 165; stripping of atmosphere by, 196

radioactive decay: creation of helium, 150–51, 153, 154; and planetary heat, 151, 193–94

rain: diamond, 181; helium, 159–62

red giant phase, 194–95

RISE (Rotation and Interior Structure Experiment), 141

rocks: asteroids as, 44; breccia, x; and challenges of exploring interior of Earth, 22–23; and comets, 44; and convection in mantle, 151–53; defined, 43; formation of, 43; and formation of planets, 29, 47, 50, 51–52; and understanding interior of Earth, 22–23, 25–26; and understanding plate tectonics, 110–13. See also meteorites; seismic forces

rocky planets, See terrestrial planets

rotation: determining rotation period of Saturn, 114–19, 135–36, 137; determining rotation period with magnetic field, 114–19; differential, 133; and formation of solar system, 35–38; of Mars, 121, 141; and moment of inertia, 142–46; and nutation, 140–41; and precession, 139–42, 144, 146; and rotational bulge, 98–100, 115–16, 136–39; of Sun and seismic forces, 133; and understanding interior of planets, 136–39

Rotation and Interior Structure Experiment (RISE), 141

Sagan, Carl, 35

Sahara and meteorites, 69–70, 75

satellites (natural), See moons

satellites and navigation, 5

Saturn: atmosphere, 160–61; core, 135, 165; dynamo, 158; formation of, 50–52, 160; as gas giant, 52; helium in, 156–62; helium rain, 159–62; magnetic field, 114–19, 158, 161–62, 164–65; metallic hydrogen in, 156–59; missions to, 114, 117, 118–19, 134, 136, 163–64, 177; and rotational bulge, 137; rotation period, 114–19, 135–36, 137; seismic forces on rings, 133–36; winds, 164–65

Saturn, moons of: number of, 45. See also Enceladus; Mimas; Titan

Schrödinger Crater (Moon), 129

secondary waves, See S-waves

seismic forces: on Earth, 122–25; Mars, 121, 126, 129–32; measuring, 122–24, 125–26; on Moon, 126, 127–29; on other planets and moons, 121, 125–36; and understanding planetary interiors, 91, 122–26 seismometers, 121, 122–23, 125–26, 127–32

self-gravity, 101

70 Ophiuchi, 56

shear waves, *See* S-waves

Shepard, Alan, 127

silicates: and formation of planets, 39, 182; melting on Moon, 187; and meteorites, 73, 75, 76, 77; on Triton, 178

silicon, 75, 182, 183, 195

16 Psyche, 78–81, 184–85, 202–3

skating, 36–38

solar flares, 12

solar system: age of, 74; boundary of, 16; formation of, 30, 33–42; leftovers, 44, 45, 50, 54–55, 63–65; and missing fifth giant, 52–53; and understanding Earth, 30–31

solar wind: and aurora borealis, 2, 6, 9–13; magnetic field of Earth as shield from, 3, 6, 9–11, 31, 114, 211; and stripping of Mars' atmosphere, 201, 211

sound and aurora borealis, 7

Space Telescope Science Institute, 38–39

spectrometers, 80, 205

spheres: deformation from Sun's gravity, 102–3; Earth as imperfect, 14, 98–100

spin: and angular momentum, 35–38, 115, 139; and dynamos, 3; and nutation, 140–41; and precession, 139–42, 144, 146. *See also* rotation

stagnant lid convection, 112

Stardust mission, 43

stars: carbon to oxygen ratio in star systems, 182–83; death of, 33, 193; defined, 46; observing births of with telescopes, 38–39, 198–202; and supernovas, 33, 193; white dwarf stars, 191–92, 194–95

Steinbeck, John, 18

stellar nurseries, 198

stony iron meteorites, 73, 76–77

structural integrity as challenge in exploration, 23

subduction zones, 98, 108, 113, 125, 153–55, 184

Sudbury Crater, x

sulfur, 79, 107–8, 210

Sun: and aurora borealis, 2, 6, 9–13; composition of compared to meteorites, 74–75; coronal mass ejections, 11–13; death of, 194–95; formation of, 34, 35, 39; gravity of, 101–3; helium in, 160–61; magnetic field of, 16, 133; Parker Solar Probe, 16; and precession of Earth, 140; proto-Sun, 34, 35, 39; proximity of Neptune and Uranus to on formation, 52; seismic forces on, 132–33; as star, 46; sunspots, 133; and tides, 101–3

sunspots, 133

super-Earths, 46, 60, 61, 195

superionic water, 172–73

supernovas, 33, 193

surface pressure, 31–32

S-waves, 123, 124

telescopes: and future of planetary
exploration, 197–202; and observation
of formation of star systems, 38–39;
and transit method of detecting
exoplanets, 59–60

temperature: buoyancy and convection,
4–5, 125, 151; in core of Earth, 28;
and diamonds, 86; as exploration
challenge, 21, 23; and formation of
exoplanets, 58; and formation of
giant planets, 50–51; and formation
of solar system, 39; and helium rain,
159–60; in mantle of Earth, 26–27;
and planetary differentiation, 47;
and plate tectonics, 113–14. *See also*
heat

terrestrial planets: comparison of, 42,
47–50; defined, 45; determining
density of, 96–97; determining rotation period, 115; exoplanets, 60;
formation of, 42–50; and rotational
bulge, 138; and volcanism, 184, 210–11.
See also Earth; Mars; Mercury; Venus

TESS (Transiting Exoplanet Survey
Satellite) mission, 195–96

Tharp, Marie, 109–10

Tharsis Mons (Mars), 101

Theia, 45, 48–49

thorium, 150–51

tides: and gravity, 101–5; and Moon's
rotational speed and distance from
Earth, 138–39; and Pluto and Charon,
179; and seismic activity on other
planets, 127, 128

Titan, 31–32, 55, 126, 176, 206, 215

TOI-4329b (exoplanet), 196

Transiting Exoplanet Survey Satellite
(TESS) mission, 195–96

transit method of detecting exoplanets,
59–60

TRAPPIST-1 system, 60, 200, 202

trenches, mapping, 110

Triton, 168, 178, 187

Trojan asteroids, 204–5

uranium, 86, 150–51, 194

Uranus: dynamo, 169–72; formation of,
50–52; liquid diamond in, 181–82;
magnetic field, 167, 169; magnetic
poles, 162; phases of water and ices
on, 167–73; proximity to Sun when
formed, 52

Uranus, moons of, *See* Ariel; Miranda

van de Kamp, Peter, 56

van Maanen's Star, 191–92

van Raadshooven, Mr., xv

Vega, 140

Venus: atmosphere of, 31–32, 146–47, 180,
207, 211; composition of, 42; core, 32,
46, 146–47; crust, 25; determining mass

of, 96, 146; dynamo, lack of, 32, 146–47;
as habitable, 215; magnetic field, lack
of, 211; mantle, 42, 207; meteorites
from on the Moon, 72–73; missions to,
187, 201, 206–7; moment of inertia,
146; precession of, 146; research
challenges of, 146–47; seismic forces
on, 126; as terrestrial planet, 42, 47;
volcanism on, 184, 186–87, 207; and
water, 201

VERITAS (Venus Emissivity, Radio
science, InSAR, Topography, and
Spectroscopy), 206–7

Vesta, 70, 72, 75

Viking 1 mission, 129

Viking 2 mission, 120

viscous heating, 160

volcanism: on Ceres, 187–88; cryovolcanism, 177, 178, 184, 187–88; on Earth,
151–52, 184, 186; on exoplanets, 210;
ferrovolcanism, 79–80, 184–85, 203;
on Io, 103–5, 184; on Mars, 187, 203; on
Moon, 187; mud volcanoes, 184, 188;
on 16 Psyche, 79–80, 184–85, 203; and
tidal warming, 103–5; on Venus, 184,
186–87, 207

Voyager 1 mission, 16, 104–5, 117, 136

Voyager 2 mission, 117, 136, 178, 187

Walker, Gordon, 56–57

Walsh, Don, 110

Washita County, Oklahoma borehole, 19

WASP-39b (exoplanet), 210

WASP-96b (exoplanet), 202

water: in boundary of upper and lower
mantle (Earth), 87, 88–89; Callisto,
174, 175; Ceres, 179; Enceladus, 45,
168, 177; Europa, 45, 168, 174–75, 205,
206; and exoplanets, 60–61, 202; and
formation of planets, 39, 51; Ganymede, 174, 175, 205; and giant planets,
51, 97, 167–69; Io, 174, 175; ionic water,
169–71; and life on other planets,
174, 180; Mars, 214; Mimas, 178; Moon,
68; phases of on Neptune and Uranus, 167–73; and plate tectonics, 113;
Pluto, 178–79; search for off of Earth,
173–79, 205–6; superionic water,
172–73; Titan, 176; Venus, 201; water
cycle, 89

waves, seismic, *See* seismic forces

Weaver, Hal, 63–64

Wegener, Alfred, 109, 111

Weiss, Benjamin, 78–79, 80, 81

white dwarf stars, 191–92, 194–95

winds on Jupiter and Saturn, 164–65

World Magnetic Model, 5

Yang, Stephenson, 57

Young, John, x

zircons, 85–86, 87

JOHNS HOPKINS
WAVELENGTHS

JESSICA FANZO, PhD

Can Fixing Dinner
Fix the Planet?

WAVELENGTHS

CAN FIXING DINNER FIX THE PLANET?
Jessica Fanzo, PhD

LISA COOPER, MD, MPH

Why Are
Health Disparities
Everyone's Problem?

WAVELENGTHS

WHY ARE HEALTH DISPARITIES EVERYONE'S PROBLEM?
Lisa Cooper, MD, MPH

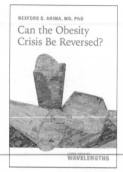

REXFORD S. AHIMA, MD, PhD

Can the Obesity
Crisis Be Reversed?

WAVELENGTHS

CAN THE OBESITY CRISIS BE REVERSED?
Rexford S. Ahima, MD, PhD

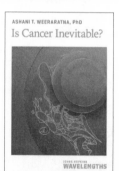

ASHANI T. WEERARATNA, PhD

Is Cancer Inevitable?

WAVELENGTHS

IS CANCER INEVITABLE?
Ashani T. Weeraratna, PhD
with Tim Wendel

RAMA CHELLAPPA, PhD with ERIC NIILER

Can We Trust AI?

WAVELENGTHS

CAN WE TRUST AI?
Rama Chellappa, PhD
with Eric Niiler

ARTURO CASADEVALL, MD, PhD
WITH STEPHANIE DESMON, MA

What If Fungi Win?

WAVELENGTHS

WHAT IF FUNGI WIN?
Arturo Casadevall, MD, PhD
with Stephanie Desmon, MA

COMING SPRING 2024

JOHNS HOPKINS
UNIVERSITY PRESS

PRESS.JHU.EDU